THE ELECTRICAL CONTRACTOR'S HANDBOOK OF CLAIMS AVOIDANCE AND MANAGEMENT

About The Authors

Harvey L. Kornbluh

Harvey L. Kornbluh is President of Construction Consultants International Corp. and is a nationally recognized construction consultant. He has had broad experience in all phases of the construction industry and particularly in the preparation and documentation of construction claims.

For over a decade, Mr. Kornbluh has been engaged in construction consulting work for various clients, including owners, architects, engineers, sureties, general, mechanical and electrical contractors and their counsel. He has been involved in claim analysis, documentation, damages analysis, estimating, and overall construction claim preparation. His work has also included preparation of sophisticated scheduling analyses using such scheduling techniques as CPM and PERT and claim presentations.

As a construction consultant, Mr. Kornbluh has assisted various trial counsel in the negotiation, trial, and arbitration of construction contract disputes. Mr. Kornbluh has also appeared as an expert witness in State and Federal Courts, and in arbitration hearings.

Mr. Kornbluh has held various managerial positions in the electrical construction industry rising from Project Engineer to Branch Manager. His duties have included overall responsibility for various multimillion dollar electrical construction projects.

Peter M. D'Ambrosio

Mr. D'Ambrosio is Vice President and General Counsel of Construction Consultants International Corp. He has represented owners, contractors, architects, engineers and sureties concerning construction claims both in the United States and overseas.

Mr. D'Ambrosio received both his Bachelor of Arts and Law Degrees from Georgetown University, Washington, D.C. He also served as an editor of the Georgetown international law journal, *Law & Policy in International Business.*

Mr. D'Ambrosio has also had a broad range of experience as a private construction attorney. As a partner in a nationally-known construction law firm, Mr. D'Ambrosio represented owners, prime and subcontractors, architects, engineers and sureties in the negotiation, arbitration and litigation of construction claims. His experience included claims concerning wastewater treatment plant, office building, hospital, condominium, hotel, school and power plant projects.

Prior to joining CCIC, Mr. D'Ambrosio was a contract manager for Westinghouse Nuclear International, a division of Westinghouse Electric Corporation which is engaged in the design, supply and overall management of power generation projects outside of the United States. Mr. D'Ambrosio was responsible for the development and implementation of risk management and claims analysis systems and for the preparation and review of contracts and scheduling data. He was also involved in the preparation and negotiation of claims arising from international construction projects.

THE ELECTRICAL CONTRACTOR'S HANDBOOK OF CLAIMS AVOIDANCE AND MANAGEMENT

Harvey L. Kornbluh
Peter M. D'Ambrosio

RESTON PUBLISHING COMPANY, INC.
A Prentice-Hall Company
Reston, Virginia

Library of Congress Cataloging in Publication Data

Kornbluh, Harvey L.
The electrical contractor's handbook of claims avoidance and management.

1. Electric contracting—United States—Management.
2. Claims. I. D'Ambrosio, Peter M. II. Title.
TK441.K67 1985 621.319'24'068 84-6757
ISBN 0-8359-1612-X

"This publication is designed to provide accurate and authoritative information in regard to the subject matter covered. It is sold with the understanding that the publisher is not engaged in rendering legal, accounting or other professional service. If legal advice or other expert assistance is required, the services of a competent professional should be sought."

From a Declaration of Principles jointly adapted by a Committee of the American Bar Association and a Committee of Publishers and Associations.

A Prentice-Hall Company
11480 Sunset Hills Road
Reston, Virginia 22090

10 9 8 7 6 5 4 3 2 1

Printed in the United States of America

Contents

CHAPTER THREE

Job Cost Control 24

CHAPTER FOUR

Contract Administration 66

CHAPTER FIVE

Scheduling and the Use of Scheduling Techniques 125

Foreword

While the role that the electrical contractor plays in the overall construction process has not changed materially over the years, the environment in which he has to operate and perform has changed significantly. It is not sufficient any more for the electrical contractor to merely bid work and then blindly rely on the other project members who include owners, architects, engineers, general contractors and other subcontractors, to dictate and control the execution of that work. The old adages that a project "will cost what it costs" and "will be completed when it is done" refer to an approach in construction administration that has long since disappeared.

The electrical contractor (as well as other specialty contractors) stands in a unique position in relation to the other project participants. Today's trend is for more of a project's workscope to be executed by specialty contractors. As such, the emphasis is on the specialty contractor to establish and maintain stronger and more reliable procedures and controls to administer the execution of its workscope. The ultimate goals of the electrical contractor are to maximize its profit while minimizing its risk and exposure. To accomplish these goals it is imperative that the contractor initially establish strong and viable work plans, budgets, and administrative procedures.

The authors have given the reader in Chapters One through Six a logical and practical plan to meet these goals. The suggestions and recommendations that are enumerated, along with the use of the sample forms that are provided, establish a practical plan to administer and monitor a project's progress. When a project that has these controls runs into problems, then the contractor will be in a much better position to recognize the impacts earlier and adjust his planning to accommodate the situation and notify other project participants.

Regardless of the commitment and ability of the electrical contractor to establish and maintain methods of adequately controlling his workscope, there are situations created by others on the project that create and justify claims. If the electrical contractor finds himself in such a situation and *if* he has previously established a solid basis for controlling and monitoring *his* workscope as suggested by the authors in Chapters One through Six, then the preparation and presentation process of a valid claim can be accomplished with a minimal amount of additional work. The added cost required to establish and maintain proper administration of a project is insignificant compared to the burden and cost that will be required to establish the necessary data "after the fact." Chapters Seven and Eight of the text clearly indicate the necessity for adequate project administration and the use of these records in the claims process.

A cursory reading of the topics addressed in the text would lead one to believe that this book was written primarily for those electrical contractors initially getting started in the industry. However, the subject matter is relevant not only to those type firms but also to the successful, established firms who have been in the industry for years. Perhaps it is for these firms that this book has its deepest impact. A changing construction environment coupled with a tightening of the economic market require a reevaluation of established procedures and controls. The adoption of the suggestions and recommendations that have been put forward by the authors will result in realized benefits.

MITCHELL W. BECKER
Contracts Manager
Dynalectric Company

Acknowledgments

The authors would like to extend their gratitude, with special thanks, to the following people for their help and assistance in assembling this handbook: Kathleen G. Coleman and Phillip R. McDonald for their editorial assistance; Robert C. Mallasch for his skill and creativitiy in design; Richard W. Lamb for his valuable critiques; and Sheryl L. Brubaker for her ability and willingness to "put it all together."

CHAPTER ONE

Introduction

In the world of construction today, you, as an electrical contractor, must have many areas of knowledge and competence in order to be successful. You must be skilled in the trade as well as able to manage your company, from the top executives to the skilled labor that works on the site. The corporation organizational structure can be an essential element in the success of your company. Individuals have to work together and contribute their expertise to make your company profitable and successful. The same techniques apply to the "success" of a construction project. All trades must work together and provide their expertise in order to build the project properly.

The Electrical Contractor's Handbook was assembled for the express purpose of aiding the electrical contractor in the management of a corporation, the costs and projects. It also provides practical solutions to problems in areas of job mobilization, scheduling, contract administration, and claims. This book discusses the proper documentation to maintain in order to avoid a claim, what documentation is needed to put a claim package together and, finally, how to negotiate, litigate or arbitrate a claim. It was written for you and your corporation to provide insight into specific areas of concern among contractors, namely, excessive costs and inefficiency. To control and reduce inefficiency and cost, we have given you a variety of forms and methods with which to document the work and expense of the construction project. Through use of these forms and methods, you will have a system of reducing cost and inefficiency before they cause damage to your project and profitability. You will also have a fully documented claim in order to support, prove, and recover losses.

As an introduction to the handbook, we would like to briefly discuss some of the problem areas for the electrical contractor which the book will cover in detail. Critical areas of concern involve scheduling techniques, cost reporting and accounting methods, and project record-keeping and documentation. Failure to address these problem areas can contribute to increased direct and indirect costs. As a result, your company becomes less competitive in the marketplace, and growth and profit decrease. What makes one contractor a success and another one a failure lies in project control, since the basics of construction, i.e., labor, construction materials, suppliers and equipment, are common factors in all projects. So it stands to reason that the contractor who exercises the greatest control over his resources and who performs his work on time will be the one who is most successful and profitable.

Increased Job Control through Improved Scheduling Techniques

In order to maintain job control through the life of a construction project, it is critical that proper management techniques, including scheduling, are utilized. The use of a suitable and proper scheduling method is more important in the 1980s than ever before. This concern is caused by such factors as spiraling inflation, improved scheduling techniques including use of computers, and increasingly sophisticated methods of claim preparation.

Unlike the situation just a few decades ago, today the more sophisticated methods of scheduling, such as the critical path method, are widely used and totally accepted, not only as tools for guiding performance during the life of the contract, but also as appropriate methods for determining the accuracy of claims made as a result of performance. When properly used, the project schedule can be a critical tool for project management in exercising control of overall job performance.

Many contractors operating today have not established or adopted a standard operating procedure for the scheduling process as a whole, which they should do. Construction planning and scheduling vary from project to project and from one office to another. In some organizations, the specifications of a particular project dictate what type of scheduling to use. This can vary the contractor's control efforts from a simple bar graph to a computerized critical path method schedule.

Construction planning and scheduling policies should not be considered separate and distinct elements from corporate policy, varying from job to job as the specifications dictate. Rather, they should be company instilled policies which all key staff members know and abide by on every construction project. Job superintendents as well as key department heads and personnel should understand and comply with these policies. Key personnel should have a clear and concise understanding of their role and responsibility with respect to each construction project. After all, it takes an entire organization, both field and office, to produce a successful project.

Emphasis on Establishment of Accurate Cost Accounting

The problem of accurate and timely cost accounting has always been a major concern to contractors. It is only through an accurate accounting cost code procedure, compatible with sound scheduling techniques, that you are able to review each project individually to determine: (1) if the costs are staying within budget; (2) that projected expenditures are based upon the amount of work left to complete, which should be defined in some type of quantitative measure along with a unit price history; and (3) most important, the time involved in its performance on the construction project.

In order to have an effective cost control program, the costs incurred and committed must be determined as soon as possible and an immediate comparison made with the anticipated progress of the plan. This point cannot be emphasized too strongly since lack of production and timely cost control procedures are key factors in the failure of construction companies today. Basically, the contractor is faced with an ever-changing environment surrounding his construction work coupled with a high dependence on labor productivity. These factors create the need for the early recognition of inefficiencies and deviations from planned progress since both money and time can quickly slip away if corrective action is not taken in a timely manner.

Record-Keeping and Documentation Are Often Neglected

In every construction organization, both large and small, a policy must be established to keep an accurate and complete documentation of every project from the prebid or estimating stage to the project completion. Documentation will serve to substantiate what occurred during the life of the project.

The detail and general character of such records, and the cost to keep them, depend largely on the kind and complexity of the contract work involved. However, it must be understood that construction claims can be virtually impossible to support unless you have a proper system to document your projects. Further, your field personnel must be properly trained to thoroughly understand what, how, and when to document the files. In many cases, field personnel do not understand what records are important, and you may end up with additional reimbursable expenditures, but lack supporting evidence to prove your case.

Increasing Trend toward Construction Claims

The volatile and highly competitive nature of the construction industry today has promoted greater awareness of claims among subcontractors, general contractors, owners, architect/engineers, and surety companies. In this period of spiraling inflation and high interest rates, parties to a construction contract are becoming increasingly aggressive in protecting and enforcing their respective rights and remedies under the contract with resultant increases in disputes and potential claims.

Contractors are also becoming very conscious of excessive allocations of their money. In the pursuit or defense of a claim, a costly item is attorney's fees. Therefore, there has been an increasing tendency toward the use of consultants rather than attorneys for the preparation and compilation of the claim package for negotiation. The decision by contractors to engage a construction consultant is based principally on the feeling among contractors that the employment of an expert at an early stage of the claims procedure can provide the detailed information necessary to avoid lengthy and expensive claim litigation or arbitration. As will be discussed in more detail in chapter seven of this handbook, it is important for you to give high priority to engaging a claims consultant early in the construction dispute since he can be of great assistance in identifying the strengths and weaknesses of both your position and that of your adversary. He can also assist in the documentation of the factual aspects of the dispute, including establishment of proper record-keeping procedures for such important areas of the project as scheduling and cost control. The attorney's specialty is in the legal field. His knowledge and skill should be used when it becomes necessary, and that is at the time of the trial or arbitration.

With greater frequency, parties involved in a construction claim are recognizing and utilizing negotiation as the most efficient and effective method of claim resolution. This trend toward negotiation of claims has been fostered by three principal forces: high costs, rising inflation, and interest. In addition to the substantial direct and indirect costs and the significant potential for delay associated with formal litigation of construction disputes, there is also a great degree of uncertainty associated with the resolution of claims through litigation, arbitration or administrative hearing. Therefore, it is beneficial to parties involved in a construction claim to avoid litigation and to pursue a dispute resolution method which offers greater probability of an outcome mutually satisfactory to both parties. Even an unsuccessful conclusion to settlement negotiations can leave you with a great deal accomplished. For example, even an unsuccessful negotiation procedure can offer the opportunity to gain information which otherwise would be unavailable until after the commencement of litigation. Additionally, this occasion is also a good opportunity to reevaluate your position and help you to decide whether to proceed with arbitration or litigation.

Recognizing the trend toward greater claims awareness in construction disputes, this handbook will also discuss two important elements of the claim resolution process: the selection of a dispute resolution procedure which will be most advantageous and effective for your claim presentation or defense, and strategies and techniques for effective claim prosecution.

CHAPTER TWO

Electrical Contractor's Project and Management Organization

Introduction

The principal goal of your organization is to plan, monitor and control the activities associated with the construction of projects to obtain the optimum combination of economy of operation and efficiency of time. To implement this objective, you must establish an organizational structure that is peculiarly suited to your company's mode of operation. Your management organization must ideally be sufficiently stable to assure action but also adaptable.

Of course, no one project organization could possibly be appropriate for every construction firm. An organizational pattern for a multinational electrical construction firm will not likely fit the needs of a small electrical contractor specializing solely in underground construction. Therefore, you must develop a management organization that is best suited to your individual unique requirements. There are, however, several well-recognized and widely accepted principles of organization that should be reviewed and applied by any electrical contractor who is seeking to develop an efficient plan of organization for his construction company. This chapter will discuss several essential steps in the development of an effective organizational plan for your electrical construction firm.

Principles of Management Organization

ITEMIZATION OF DUTIES

Regardless of the size of your electrical construction firm, running such a company necessarily involves certain basic duties. Although these responsibilities will be carried out by a few individuals in a small organization, these same activities may be the responsibility of many personnel in a large firm. The central issue concerning the itemization of duties is the level of detail which should be the objective of such a listing. The answer to this question depends upon the size of your company and the number of employees you will utilize in the performance of the work. It is reasonable to assume, however, that the larger the company, the more detailed the itemization of duties.

For the purpose of illustrating our discussion, we are including the following list of personnel functions and responsibilities, which is not intended to be all-inclusive.

Executive

Business Organization

Management Organization

Financial Structure

Long Range Planning

Operating Procedures and Policies

Personnel Relations and Policies

Public Relations

Labor Relations

Contract Negotiation and Execution

Legal Matters

Salaries and Other Compensation

Scope of Operations

Capital Expenditures

Auditors and Audits

Banking

Purchasing

Subcontracts

Requisitions

Equipment Rental/Purchasing

Inventories

Purchase Orders

Licenses

Bonds

Releases of Lien

Building Permits

Guarantees and Warranties

Insurance (Company and Project)

Estimating

Decision to Bid

Obtain Bidding Documents

Visit the Site

Prebid Conferences

- Quantity Take-off
- Vendor Quotations
- Pricing
- Preparation of Proposal
- Bid Bond
- Submission of Proposal
- Vendor and Material Quotations

Accounting

- General Books of Account
- Billing
- Collections
- Cost Records and Reports
- Financial Reports
- Payment of Invoices
- Tax Returns and Payments
- Wage and Personnel Reports to Public Agencies
- Bank Deposits
- Payroll and Records
- Office Services (Personnel)

Engineering

- Field and Office Engineering
- Planning and Scheduling
- Cost Accounting
- Project Payment Requests
- Shop Drawing Review

Cost Reports

Labor Relations

Accident Reports

Construction

Supervision of Construction

Hiring Labor

Timekeeping

Project Cost Data

Project Accident Data

Safety Program

Project Record-keeping

Project Progress Reports

Change Orders

Equipment Maintenance

DIVISION OF DUTIES

After you have itemized the particular duties of your organization, the next phase in the development of a management organization is to subdivide these duties into categories which can be assigned to particular individuals. For example, assume that your electrical construction firm is a small partnership consisting of only two partners. The duties may be divided, with one partner to supervise office activities and the other to supervise field operations. Under this organization, the office partner would be responsible for all duties related to purchasing, estimating and accounting. The partner supervising field operations would have duties including engineering and construction.

This example clearly illustrates that an employee of a small firm will normally be assigned responsibilities of a much broader range and diversity than an employee of a larger, more diversified company. Stated differently, members of a large organization intrinsically have relatively narrower and considerably more specialized responsibilities. For example, if you are a large electrical contractor, you may need several employees to accomplish the same tasks one employee accomplishes in a small firm because each employee will be involved with a small aspect of the total activity.

STRUCTURE OF THE ORGANIZATION

Although each electrical contractor organization is unique in some respects, it is common practice in electrical construction to establish intracompany departments or jurisdictions. Under such a scheme, each department generally has equivalent authority and although interrelated, each operates relatively independently of the other departments. For example, each department is assigned a specific functional area of responsibility, such as engineering, and is supervised by an overall manager who possesses training, skill, and experience in that particular element of the electrical construction business.

Each individual department developed under such an organizational structure is then divided vertically. This procedure allows establishment of direct lines of supervision so that each individual along a particular line is accountable to the person or persons above him and acts in a supervisory capacity to those individuals below him.

In later chapters of this book which cover such topics as project management, project cost control, and contract administration, for purposes of illustration we have utilized a typical contractor organization structured according to major functional departments including contract administration, accounting and finance, estimating, and construction and projects.

Project Management

INTRODUCTION

One of the essential elements of proper project management and control is the establishment and implementation of guidelines for the responsibilities of key project personnel. In particular, your company must establish a policy concerning the responsibilities for the project manager and project (field) superintendent on each job. Important functions such as contract administration, cost control and planning, and reporting and record-keeping procedures must be included in the policy in order to maximize control and efficient use of labor, material, and equipment, and thus increase the profitability of the individual projects. This chapter will discuss the respective responsibilities of the project manager and project (field) superintendent concerning these functions and can be used as a guideline in the establishment of your own corporate policies concerning project management.

PROJECT MANAGER

The project manager is normally assigned to supervise several projects and is fully responsible for every phase of each of these projects. This arrangement may occur before the time of the initial contact with the customer[1] or as late as after award of the contract. Unless reassigned, the project manager is responsible for all operations through the time of final completion of the job and final payment to your company.

One of the initial responsibilities of the project manager may be participation in the estimating and bidding or contract negotiation stage of the project. The role of the project manager in estimating, if any, is generally established by the policies and procedures of your electrical construction company. Although recognizing the need to have the individual who is charged with the overall responsibility for a particular project involved at its inception, many larger electrical contractors have established separate estimating departments which have the principal responsibility for developing cost information for bidding or negotiation of electrical construction contracts. Under these circumstances, the project manager may be assigned to the particular project only after the electrical contractor has received notice of award for the work.

When you have received notice of contract award or notice to proceed with the project, the project manager assumes responsibility for mobilization of labor, material, and equipment as well as management of the company's resources assigned to the particular project. He is also responsible for monitoring the job through the use

---

[1]We have utilized the term *Customer* throughout the handbook to signify the party with whom the electrical contractor enters into an agreement. In practice, this customer is usually a general contractor or construction owner.

of a reporting and record-keeping system which keeps track of job costs and of the progress of the work. He must also supervise and control the overall process of documenting, pricing, and negotiating change orders with the customer. In order to perform these responsibilities in the most effective and economical manner, he must possess a complete, in depth familiarity with all aspects of the project including the contract, project cost, and construction schedule until final demobilization and final payment have occurred.

Preconstruction Responsibilities

Many electrical construction companies ordinarily involve the project manager in the preconstruction phases of a particular project. This corporate policy is based principally on the recognition of the advantage of involving the project manager in the development of a project in order to obtain the benefit of his field experience and particular knowledge of similar projects he may have managed in the past. The role of the project manager during preconstruction may be eliminated or limited, however, as a result of the existence of a central estimating department or because of a corporate policy establishing a combination of estimating and project management forces to develop the cost for an individual project.

If your company policy utilizes involvement of a project manager in the prebid or preconstruction phase of the project, the project manager's initial responsibilities arise when he is assigned to the project for which the customer is soliciting bids or negotiating contracts. At this point, the project manager makes the initial contact with the customer and, if he does not actually estimate the project, he participates to a great extent in the estimating process. The project manager, as the primary contact with the customer, then handles the bidding or negotiation of the job.

After you have successfully obtained the award of the contract from the customer, one of the initial tasks of the project manager is to issue an appropriate Project Data Report (Form No. 01) and to obtain a job number so that the project can proceed. The purpose of the Project Data Report is to provide all affected departments of your organization, such as contract administration, accounting, and planning and scheduling, with a complete report of important project information. This procedure allows you to take a coordinated and comprehensive approach to the project in order to ensure that all departments of your home office are supporting the project in a complete, efficient and timely manner.

PROJECT DATA REPORT

☐ INITIAL PROJECT

☐ MODIFICATION NO. ____________

DIVISION ____________ JOB/PROJECT WORK ORDER NO. ____________

JOB OR PROJECT NAME ____________ DATE ____________

LOCATION ____________ ORDER NO. ____________

CUSTOMER ____________

NOTE: THIS FORM MUST BE FILLED IN AND DISTRIBUTED BEFORE MATERIAL CAN BE PURCHASED OR LABOR PREFORMED. CHECKED INFORMATION MUST BE COMPLETE FOR ALL JOBS. - OTHER INFORMATION REQUIRED FOR MAJOR JOBS ONLY.

WORK INFORMATION

SCOPE OF WORK ____________

APPROXIMATE START DATE ____________ APPROXIMATE FINISH DATE ____________

LATE COMPLETION PENALTY $ ____________ PER DAY

ARE WE BONDED? ☐ YES ☐ NO BY WHOM? ____________

FIELD OFFICE ADDRESS ____________ PHONE ____________

FIELD OFFICE CONTACT ____________ POSITION ____________

SHIPPING ADDRESS ____________

INSPECTION JURISDICTION ____________ UNION JURISDICTION ____________

WHAT SECURITY IS INVOLVED? ____________

IS PROJECT CODED? ____________ IS SMOKING ALLOWED? ____________

OTHER ____________

BILLING INFORMATION

BILL TO ____________ STREET ____________

CITY ____________ PHONE ____________ OFFICE CONTACT ____________

POSITION ____________

TYPE OF JOB: ☐ BID ☐ NEGOTIATED ☐ T & M ☐ CPFF

OTHER ____________

PROGRESS BILLING ON ____________ THE DAY OF EACH MONTH, LESS ______ % RETENTION.

ARE PAYROLL AFFIDAVITS REQUIRED? ☐ YES ☐ NO NO. OF COPIES ____________

SEND PAYROLL AFFIDAVITS TO: ____________

SPECIAL INSTRUCTIONS: ____________

PERSONNEL INFORMATION

PROJECT MANAGER ____________ ESTIMATOR ENGINEER ____________

FIELD SUPERVISOR ____________ OTHER ____________

PROJECT MAIL ADDRESS ____________ CITY ____________

JOB OFFICE LOCATION ____________ GATE NO. ____________ PHONE ____________

CONTRACT INFORMATION

ESTIMATED MATERIAL & FREIGHT	$ ____________	APPROVED
SALES TAX	$ ____________	____________
SUB CONTRACTS	$ ____________	PROJECT MANAGER
TOTAL ESTIMATE MATERIAL COST	$ ____________	
DIRECT LABOR (HOURS ______)	$ ____________	____________
CRAFT SUPERVISION LABOR (HOURS ______)	$ ____________	DIVISION CONTRACTS MANAGER
TOTAL ESTIMATE LABOR COST	$ ____________	
OTHER DIRECT COSTS	$ ____________	____________
OVERHEAD (RATE ____ % OF $ ____________)	$ ____________	BRANCH/DIVISION MANAGER
TOTAL DIRECT COSTS	$ ____________	____________
ESTIMATED PROFIT/FEE (RATE ____ %)	$ ____________	CORPORATE DIRECTOR OF CONTRACTS
SELLING PRICE-THIS CONTRACT OR MOD	$ ____________	
SELLING PRICE-PRIOR CONTRACT TOTAL	$ ____________	
ADJUSTED SELLING PRICE	$ ____________	

Form No. 01: Project Data Report

In addition to preparing a Project Data Report after contract award, the project manager prepares a report each time the customer issues any modification. The information required to complete this document may be obtained from a review of the contract, the estimate, and other pertinent project records. The cost breakdown by element must agree with the estimate end sheet. Data covered in the initial Project Data Report need not be repeated when reports based on change orders are issued. The original of the report should be forwarded to the contract administration department or contracts manager, and copies should be distributed to the appropriate senior manager, such as the vice-president of operations or construction, and the project file.

The project manager must also prepare a coded breakdown of labor, material, and equipment costs based on the estimate (which will be discussed in chapter three). This breakdown should be forwarded to the accounting department or controller along with the original estimate. Utilizing the planned expenditures for labor, material, and equipment, the project manager must participate in the development of a comprehensive construction schedule which will allow him to complete the job no later than the contractually required completion date without exceeding the contract budget. As will be discussed more fully in chapter five, this project schedule, which should include procurement, engineering, and construction activities, is the project manager's principal tool for project control and management.

Between the date of the award of the contract and the start of actual construction, the project manager must also secure all licenses, permits, bonds, and insurance certificates required under the contract with the customer. The functions required for project mobilization and start up are discussed in detail in chapter six of this book.

During this time, the project manager should also become completely familiar with two key sets of documents: the bid under which the contract was awarded or negotiated and the contract documents governing performance of the work. If the project manager did not actually prepare the bid, he must become completely familiar with it before construction begins. The project manager must also become thoroughly familiar with the contract documents including purely legal issues, such as the general or special conditions to the contract, as well as the contract drawings and technical specifications which define the scope of work.

Construction

One of the initial duties of the project manager is the development of that portion of the overall construction schedule which will be performed by his company. Although the development of this schedule will be discussed in detail in chapter five, this activity initially involves the review of the requirements of the contract relating to the schedule. For example, before any effort is made to prepare the schedule, the contract requirements concerning the kind of schedule method (e.g., critical path, precedence diagram), the schedule reporting requirements including number, detail, level, and frequency of reports, and any unusual requirements concerning the review of scheduling aspects of contract modifications should be reviewed by the project manager. As part of these preliminary efforts, the project manager should also meet with the customer to discuss the relationship between the sequence and duration of activities and the overall scheduling requirements of the project.

Pursuant to these initial discussions the electrical contractor should develop, in detail, sequences and durations for the individual electrical work activities required on the project based on the manpower, material, and equipment included as part of your bid or negotiated price. This information should then be reviewed with the customer in order to coordinate the performance of this work with the work being performed by other contractors on the project.

As part of his responsibilities during the construction phase, the project manager also mobilizes the labor, material, and equipment required for the project. These activities include the responsibility for ordering material through issuance of a purchase requisition to your company's purchasing agent or department. Upon receipt of materials which have been ordered, the project manager also supervises the preparation and submission of receiving reports. After receipt of this material, the project manager also has the overall responsibility of maintaining control of the material and of disposing of surplus material to obtain maximum credit to the project.

The project manager is responsible for appointing an on-site field superintendent with the approval of the vice-president of construction or an equivalent senior manager. Under the circumstances of a union labor contract, he approves the hiring of personnel through the local hiring hall agent and maintains contact with the local union business agent. The project manager manages the construction operations performed by the forces on the job by assigning appropriate work areas, supervising construction, maintaining a sound safety program, etc.

The project manager is the person responsible for obtaining the necessary equipment to perform the work. He has overall responsibility for requisitioning necessary equipment, accepting it upon delivery, maintaining the equipment while it is on the job site, and preparing transfer notices when equipment is no longer required for the performance of the work.

Throughout the actual construction of the job, the project manager must constantly be alert for new methods of performing the work and new materials which may be used. Whenever possible, he should introduce cost improvement techniques to maximize the efficient use of labor, material, and equipment.

Changes to the Contract

The project manager is the party principally charged with handling the change order process. As is discussed more fully in chapter six, changes to the contract can occur by either formal action on the part of a customer or constructive actions or failures to act which result in modifications of the electrical contractor's scope of work. In either situation, the project manager must know and understand the provisions of the contract concerning changes and in particular he must strictly adhere to requirements for notice to the customer, submission of documentation supporting either the constructive or formal change, and other contractual responsibilities which must be carried out in order to obtain a modification of the contract. Although many of the constructive changes may ultimately be disputed by the customer with resultant claims by your company, familiarity with the changes provisions of the contract, including requirements for notice, and execution of the responsibilities under these clauses will serve to place your company in the most advantageous position in the event of a later claim with the customer. Similarly, both formal and constructive modifications of the contract require that you maintain complete and accurate documentation, including correspondence, reports, meeting minutes, photographs, etc., concerning the scope and cost of work in order to support the issuance of an appropriate contract modification or a later claim, if a dispute arises between the parties.

If the customer has issued a formal request for a change of your scope of work, the project manager has the initial assignment of preparing an estimate for the cost of the contract modification in terms of additional money as well as the responsibility of the extension of the project schedule, if appropriate. This effort should include a review of the impact of the change on the project schedule. Methods for analyzing the impact of a change on the project schedule are discussed more fully in chapter five.

Based on that estimate, the project manager must then submit a change proposal to the customer which presents the most favorable, justifiable terms. Based on that proposal, the project manager will carry out negotiations to finalize the amount of any reimbursement and/or time extension provided by the change order. As part of the change order process, the project manager must also consider the indirect costs (including extended job site and field office overhead) and other expenses which may be incurred as a result of the performance of this change. As an alternative, the project manager may reserve the company's right to later recover the indirect costs of this change, required by the circumstances and nature of the change.

As a continuation of his responsibility to maintain a thorough familiarity with the contract documents and the company's estimate, the project manager must be fully knowledgeable of the content of each change proposal and each change order actually issued. He should be aware of formal changes made to the contract price or the contract time for performance as well as the actual effects of the change on the company's costs and job progress. Based on this information, the project manager should make appropriate revisions and updates to the planned schedule for performance.

MAINTENANCE OF REPORTING AND RECORD-KEEPING SYSTEM

As part of his responsibilities during the construction phase, the project manager should maintain various financial and administrative reports which are described in detail in chapter four of this desk book. The basic administrative report which should be maintained by the project manager is the daily job diary. In his diary, the project manager should record specific key events, such as information concerning job meetings or inspections which affect the progress of the job, as well as general information regarding job progress. This diary also affords the project manager the opportunity to record, in a detailed manner, the facts of any potential claims situation, including changes, out-of-sequence, disrupted or delayed contract performance, or any other significant offense which may entitle you to additional compensation or time for contract performance. Of course, notations in the project manager's diary do not take the place of the requirement that your company carry out its contract obligations concerning notice of potential claims, including timely requests for additional monies or extensions of contract performance. This diary should also include memoranda concerning telephone conferences relating to construction progress.

The project manager must also maintain various financial reports which allow your company to monitor the financial aspects of the job. He must keep track of time worked by employees and audit and submit time sheets for the payroll department. On a broader scale, the project manager is responsible for preparing inventories of the job and completing monthly labor and material cost summaries.

The project manager must negotiate progress payment requisitions with the customer and promptly report the appropriate amount for billing to the company accounting department.

Finally, the project manager has overall responsibility for numerous miscellaneous aspects of the project's administration, reporting, and record-keeping systems such as the preparation of accident, theft, and casualty reports, and the maintenance of a complete set of marked-up contract drawings reflecting current job status. These duties are described in more detail in chapters three and four.

Ultimately, the project manager prepares and submits the job completion notice when the time has come to demobilize the labor, material, and equipment on the job as well as to complete demobilization in an expeditious manner.

Delegation of Duties

The project manager may delegate any of his functions or his authority as necessary to achieve the efficient and timely completion of the project, but he remains fully accountable for the performance of his assignment at all times. In actual practice, the project manager's delegation of functions and authority will often fall upon the project (field) superintendent. This relationship will vary from the complete delegation of authority to the project superintendent to mere assistance by the project superintendent in the project manager's tasks.

PROJECT (FIELD) SUPERINTENDENT

Under the supervision of the project manager, the project (field) superintendent is fully responsible for the successful execution of the project. He should be completely familiar with the terms and conditions under which the contract and bid were obtained. He also has the responsibility of assisting the project manager in planning for the performance of the project, including the mobilization of labor, equipment, and material required to complete the work, supervision of construction operations, and coordination with the customer's on-site representatives.

The project superintendent also has the important responsibility of compiling necessary progress reports, in order to keep the project manager fully advised of the progress and problems of the work. This information is needed so that the full capabilities of your company may be brought to bear whenever they will be helpful in the successful completion of the project.

The functions and responsibilities which primarily are those of the project (field) superintendent include:

Contract Administration

Since the project superintendent has daily contact with the customer, he should have a comprehensive and clear understanding of the scope of the work and any modifications to the contract. As part of this responsibility, the project superintendent must also be thoroughly familiar with the bid under which the project was acquired by your company including the budget for labor, equipment, and material needed to complete the project, according to its original schedule and cost.

The project superintendent will also have primary responsibility for the maintenance of all on-site project records including the project file and cost information. This record-keeping responsibility also includes the duty of documenting all express or constructive requests by the customer for new, revised or additional work beyond that included as part of the original scope of work. As will be discussed more fully in chapter seven, the project superintendent, in conjunction with the project manager, will have primary responsibility for the comprehensive and complete recognition of claims and potential claims for presentation to the customer.

Project Scheduling and Reporting

The project superintendent has the joint responsibility for the preparation of the original project schedule and its implementation in the field. The planning of your company's performance will necessarily depend upon the ability to sequence and schedule the labor, material, and equipment budgeted for this project, in a manner and method which will allow efficient and timely completion of the work. The methodology for completing this important critical assignment, which is discussed more fully in chapter five of the manual, is an essential element of proper job control and, therefore, increased productivity and profitability of your company. As the full-time on-site representative of your company, the project superintendent will also assist the project manager in maintaining and updating the project schedule. This responsibility will include the duty of revising the project schedule according to approved and constructive changes, as well as notifying the customer concerning potential liability for any increased costs or extended time of performance for the completion of work on the project.

In an effort to keep the project manager and your home office aware of the status of the project, the project superintendent must also maintain a daily job diary to record activity on the project. This duty could also be carried out by the maintenance of a marked-up set of contract drawings showing the status of work on the project at any period during contract performance.

The third important element of the project superintendent's planning and reporting responsibilities concerns the preparation of progress payment requisitions. Since the project superintendent is the on-site, full-time representative of your company, he is charged with the duty of working closely with the project manager concerning the accurate, complete, and timely preparation of requisitions for progress payments during the course of the project. Performance of this duty necessarily includes review of the work status with the project manager, as well as with representatives of the customer and architect, if appropriate. The project superintendent also has the responsibility of providing information to the customer concerning any details or discrepancies related to your contract performance.

Mobilization, Construction and Demobilization

Potentially, the project manager will delegate many of his responsibilities concerning mobilization, operation and demobilization of labor, material, and equipment to the project superintendent. This delegation is based to a great extent upon the fact that the project superintendent will have primary responsibility for the day-to-day operations at the construction site. Although these duties are discussed more fully in chapter four, the project superintendent's responsibilities concerning the on-site contract performance of your company are to:

a. Request necessary licenses, permits, bonds and insurance certificates.

b. With the approval of the project manager, hire personnel through a local hiring hall agent and prepare start and termination notices.

c. With the approval of the project manager, appoint general foremen, and foremen as required, assign work and supervise operations.

d. Maintain a constant vigil for new methods and materials, and introduce cost improvement techniques wherever and whenever possible to maximize efficient and economical use of labor, material, and equipment.

e. Keep time on employees, audit and submit time sheets and payroll or, as required, prepare on-site payroll and pay employees.

f. Order material by preparing purchase requisitions for approval of project manager.

g. Receive materials, prepare and submit receiving reports, and maintain material control documentation.

h. Prepare requisitions for necessary capital equipment for approval of project manager, accept delivery, maintain while on the job, and prepare transfer notices when equipment is no longer required on your particular project.

i. Complete demobilization of labor, material, and equipment in an expeditious manner when the project is complete.

In order to efficiently, economically, and satisfactorily perform his functions, the project superintendent may delegate any of his responsibility or authority to appropriate general foremen or foremen. Like the project manager, however, he remains fully accountable for the performance of his assignment at all times.

CHAPTER THREE

Job Cost Control

Introduction

Accurate and timely cost accounting and project status reporting have always been major concerns to contractors. It is only through an accurate accounting and cost code procedure, compatible with sound scheduling techniques, that you can review each project individually. This is necessary to determine if the costs are staying within budget; that projected expenditures are based on the accurate amount of work left to complete, defined in some type of quantitative measure along with a unit price history; and most importantly, the time involved in the work completion performance.

In order to have an effective cost control program, the costs incurred and committed must be determined as soon as possible and an immediate comparison made with the planned expenditures. This point cannot be emphasized too strongly since both lack of production and lack of timely cost control procedures are key factors in the failure of construction companies today. Basically, you are faced with an ever-changing environment surrounding your construction work coupled with a high dependence on labor productivity. These factors create the need for the early recognition of inefficiencies and deviations from planned progress since both money and time can quickly slip away if corrective action is not taken in a timely manner.

The control of the financial situation within your company begins with the establishment of a budget for each construction project. The budget is used in the initial phases of the project and its importance increases as the project continues. Closely associated with the development of the budget is the assignment of job numbers to each of the various projects. Job numbers play an important role in maintaining organization within the financial department of your company. To assist in the classification of a particular project, you should implement a standard job cost code for each job.

This chapter will also discuss important elements of job control including contract funding, material commitments, the daily labor distribution and insurance procedures.

Establishment of Budgets, Job Numbers and the Standard Job Cost Code

ESTABLISHMENT OF BUDGETS AND JOB NUMBERS

Organization of costs and labor will improve your efficiency as well as increase both your productivity and profitability. This section outlines methods which have successfully been used in establishing direct job cost budgets, job numbers, and standard job cost coding for electrical construction projects.

THE PROJECT BUDGET AND COST CONTROL

In the initial phase of each construction project, the primary consideration is the budget. It should be established for every electrical construction job as it is a vital tool for the cost control of each project.

Budgets are proposed by the project manager to meet the approval of the area manager or equivalent supervisor *prior to the is-*

suance of a job number. The initial development of the budget should be the first concern of the project manager. The use of the Budget Request form (see Form No. 02) is essential in establishing budgets and assigning the standard job cost code.

Budgets may be established at any one of three levels. The simplest budget breakdown is referred to as the Major Cost Level. It is a breakdown of the budget into three categories: production costs, subcontracts, and other direct job costs.

The budget with the next level of detail is entitled the Intermediate Breakdown Level (see page 28). This analysis combines the more general Major Cost Level breakdown with various groups and categories of cost under each major classification. The Intermediate Breakdown gives a more detailed analysis of the prospective budget.

The Detailed Breakdown Code (page 29) is the most sophisticated and comprehensive job cost analysis. It has the potential to be extremely discriminating in the application and segregation of project funds. This analysis is valuable for an accurate accounting of individual cost items and also is useful for monitoring and controlling the growth and development of the project. This cost format is also helpful in establishing a category of cost overrun and associating it with a specific item or group of items.

The necessary level of detail required for each construction project should be determined by the manager who is knowledgeable about the project. The development of a cost breakdown can be accomplished using either the original contract amount or project manpower assignments, depending upon the specialized requirements of your company.

Budgets must also be revised as change orders are negotiated and executed. When it is necessary, for schedule or cost reasons, to proceed on a change prior to negotiation of price, a preliminary budget adjustment should be processed, with the final adjustment being submitted after the completion of price negotiation. Approval or upper management may be required if the budget will exceed firm negotiated contract costs.

IMPLEMENTATION AND USE OF THE PROJECT BUDGET

Upon receipt of a contract or notice to proceed on a job, the project manager prepares the Budget Request, indicating the level of detail for monitoring and accumulating job costs by entering budget figures at the cost breakdown level desired. In the case of a change

BUDGET REQUEST
ELECTRICAL CONSTRUCTION

JOB NAME: ______________________ DIVISION CONTROLLER ____________

JOB NO. ____________ CHANGE NO. ____________

PRELIMINARY: ____________ FINAL: ____________ DATE: ____________

START DATE: ____________ COMPLETION DATE: ____________

MAJOR AND IMMEDIATE JOB COST CODES	LABOR HOURS	LABOR DOLLARS	MATERIAL DOLLARS	TOTAL DOLLARS
10-100 CONDUIT & RACEWAYS				X
-200 WIRE & CABLE				X
-300 DISTRIBUTION EQUIPMENT				X
-400 FINISH WORK				X
-500 SPECIAL SERVICE SYSTEMS				X
-600 MOTORS & CONTROLS				X
-700 GROUNDING & BONDING				X
-800 OUTSIDE DISTRIBUTION				X
-900 OTHER PRODUCTIVE COSTS				X
				X
				X
				X
10-XXX TOTAL PRODUCTIVE COSTS				
30-XXX SUBCONTRACT COSTS				
40-100 SUPERVISORY & ADMIN. SALARIES (NON-CRAFT)			X	X
-200 PAYROLL TAXES			X	X
-300 EQUIPMENT COSTS		X		X
-400 SMALL TOOLS & SUPPLIES		X		X
-500 ALL OTHER MISCELLANEOUS COSTS		X		X
40-XXX TOTAL OTHER DIRECT JOB COSTS				
TOTAL DIRECT COSTS				
INFORMATION ONLY: OVERHEAD AT: ______ %				
PROFIT AT: ______ %				
TOTAL CONTRACT PRICE				

PROJECT MANAGER ____________ BRANCH MANAGER ____________ DIVISION MANAGER ____________

Form No. 02: Budget Request

Electrical Construction Standard Job Cost Code Major and Intermediate Breakdown Level

10-000	**PRODUCTIVE COSTS**
10-100	Conduit & Raceway
10-200	Wire & Cable
10-300	Distribution Equipment
10-400	Finish Work
10-500	Special Service Systems
10-600	Motors & Controls
10-700	Grounding, Bonding
10-800	Outside Distribution Systems
10-900	Other Productive Costs
20-000	**UNASSIGNED (Reserved for Expansion of Production Costs)**
30-000	**SUBCONTRACTS**
40-000	**OTHER DIRECT JOB COSTS**
40-100	Supervision and Support Labor (non-craft)
40-200	Payroll Taxes, Insurance and Fringe Benefits
40-300	Equipment
40-400	Small Tools and Supplies
40-500	Miscellaneous Expenses

order, the previously approved Budget Request is amended to take into account the change. If the budget is being developed prior to negotiation of price, the block entitled "preliminary" is checked. If a firm price has been established, the block entitled "final" is checked. A copy of the contract letter of intent or other customer documentation authorizing your company to proceed should be attached to the budget request. In addition, all preliminary budget requests require supporting documentation to justify the necessity to proceed prior to receipt of customer funding.

The fully completed Budget Request, together with supporting documentation, is presented to the area manager for his approval or for forwarding to higher management. The approved request is forwarded to the responsible controller or manager of finance, depending upon the size and organizational structure of your company.

Electrical Construction Standard Job Cost Code
Detailed Breakdown Code Description

10-000	PRODUCTION COSTS
10-100	**CONDUIT & RACEWAY**
10-101	Exposed Steel Conduit
10-102	Exposed Aluminum Conduit
10-103	Buried Steel Conduit
10-104	Electrical Metallic Tubing (EMT)
10-105	Nonmetallic Conduit & Fittings
10-106	Conduit Fittings & Accessories
10-107	Sheet Metal Outlet Boxes
10-108	Cast Metal Boxes & Accessories
10-109	Underfloor Ducts & Accessories
10-110	Cable Trays & Supports
10-111	Conduit & Box Supports
10-112	Mandrel & Clean Conduits
10-113	Wiremold & Fittings
10-114	Pull Wire
10-115	Wire Ways, with Supports and Hangers
10-200	**WIRE & CABLE**
10-201	Single Conductor Wire
10-202	Multiconductor Cable
10-203	Instrumentation Cable
10-204	Hi-Voltage Cable
10-205	Direct Burial Cable
10-206	Splicing
10-207	Terminations
10-208	Cable Supports
10-209	Aerial Conductor
10-300	**DISTRIBUTION EQUIPMENT**
10-301	Main Switchgear
10-302	Unit Substations
10-303	Distribution Panels
10-304	Lighting Panelboards
10-305	Power Panelboards
10-306	Transformers
10-307	Motor Control Centers
10-308	Bus Duct & Supports
10-309	Fused Cutouts
10-310	Disconnect Switches

Standard Job Cost Code and Description (cont.)

10-400	**FINISH WORK**
10-401	Wiring Devices, Receptacles & Switches
10-402	Lighting Fixtures & Accessories
10-403	Florescent Fixtures & Accessories
10-404	Incandescent Fixtures & Accessories
10-405	Lamps
10-406	Ballasts
10-407	Lighting Standards or Poles
10-408	Plates & Covers
10-409	Special Fixture Hanger Systems
10-500	**SPECIAL SERVICE SYSTEMS**
10-501	Telephone System
10-502	Heating & Air Conditioning
10-503	Fire Alarm & Heat Detection
10-504	Nurses & Doctor Call
10-505	Sound & Paging
10-506	Television
10-507	Clocks
10-600	**MOTORS & CONTROLS**
10-601	Set Motors
10-602	Connect Motors (Includes Flex & Fittings)
10-603	Pushbutton Controls
10-604	Pilot Devices
10-605	Starters & Disconnect Switches
10-606	Junction & Pull Boxes
10-607	Testing
10-700	**GROUND & BONDING**
10-701	Ground Conductor
10-702	Ground Rods, Wells, Plates
10-703	Cadweld Connections
10-704	Lightning Protection System
10-705	Structure Bonding
10-706	Lightning Arrestors
10-707	Cathodic Protection System
10-708	Shallow Trenching
10-800	**OUTSIDE DISTRIBUTION SYSTEMS**
10-801	Trenching, Excavation & Backfill
10-802	Concrete Encasement, Footing, Slabs

Standard Job Cost Code and Description (cont.)

10-803	Unit Substations
10-804	Steel Erection
10-805	Poles & Accessory Hardware
10-806	Transformers
10-807	Oil Switches
10-808	Aerial Conductors
10-809	Fencing
10-810	Manholes
10-811	Concrete Pull Boxes
10-812	Raceways
10-900	**OTHER PRODUCTIVE COSTS**
10-901	Mobilization
10-902	Temporary Power & Lighting
10-903	Material Handling
10-904	Material Warehousing
10-905	Demolition
10-906	Craft Supervision
10-907	Testing
10-908	Welding
10-909	Cleanup

NOTE: The preceding code sections no doubt do not include all material items the estimator will list on take-off. When additions occur, include these items under the code number most closely associated with the items. For example, Marking Cable and Hypod Test are to be classified under Code 200—Wire & Cable—and assigned an identification number such as 210 and 211.

40-000 OTHER DIRECT JOB COSTS

40-100	**SUPERVISION AND SUPPORT LABOR**	
	-101	Supervision (noncraft)
	-102	Engineering
	-103	Clerical
	-104	Warehousing
40-200	**PAYROLL TAXES, INSURANCE & FRINGE BENEFITS**	
	-201	Payroll Taxes
	-202	Workmens Compensation Insurance

Standard Job Cost Code and Description (cont.)

	-203	Public Liability & Property Damage Insurance
	-204	Union Fringe Benefits
	-205	Subsistence & Mileage
40-300	**EQUIPMENT**	
	-301	Depreciation & Lease Payments (Includes Rentals)
	-302	Maintenance, Repairs, Fuel & Operations Costs
40-400	**SMALL TOOLS AND SUPPLIES**	
40-500	**MISCELLANEOUS EXPENSES**	
	-501	Telephone and Telegraph
	-502	Utilities
	-503	Services—Janitorial, Drinking Water, etc.
	-504	Office Supplies & Postage
	-505	Reproduction & Drafting Services
	-506	Office Rentals
	-507	Office Machine Rentals
	-508	Licenses & Permits
	-509	Performance & Bid Bonds
	-510	Legal & Audit
	-511	Insurance Other Than Payroll
	-512	Transportation Costs—Equipment Only
	-513	Outside Engineering & Consulting Services
	-514	Travel & Business Conference Expenses
	-515	Other

The division controller, or the department of finance, assigns the job number and the standard job cost code, which will be discussed in the next part of this section, and returns a copy to the project manager and the office manager.

Where cost data is required at the Detail Cost Level, an exhibit must be attached to the completed Budget Request outlining the additional detail required by setting forth the standard job cost codes to be utilized within each intermediate breakdown level. A budget must be provided for each detail breakdown required.

THE IMPORTANCE AND USE OF THE STANDARD JOB COST CODE FOR ELECTRICAL CONSTRUCTION

The standard job cost code provides the structure from which a cost breakdown can be developed and implemented toward the establishment of budgets and the accumulation of actual costs for all electrical construction jobs. The computerization of the job cost code is advantageous for efficient and economical monitoring of the job budget and periodic job cost information.

The standard job cost code list sets forth the detailed cost account structure to be utilized by your company's construction, project or operations group for the organization and development of the budget. The ability to accumulate direct job costs for all electrical jobs is an additional function and benefit of the standard job cost code list.

The structure of the account code consists of three series of numbers. The first set of numbers consists of two digits and is designed to distinguish the various types of costs: Production Costs (10 and 20, if needed); Subcontract and Supplier Costs (30); and Other Direct Job Costs (40).

Associated with each two-digit identification code is a three-digit dash number which is used for the purpose of specifying an activity within the respective codes. For example: Under 10-000 Production Costs are -100—Conduit and Raceway, and -200—Wire and Cable. The three-digit dash number is associated with the Productive Cost series and the Other Direct Job Costs section, but is not utilized in the Subcontract Costs category.

An additional two-digit cost breakdown number is added to the code in order to attribute a specific labor hour and labor cost to an item (see page 29). The Cost Breakdown Assignment Memo is necessary for the development of a working budget and will serve as a method to monitor cost during the duration of a project.

The cost breakdown analysis is a form which separates the construction project into several different classifications. The initial division is geographical or a division by area or floor, depending on the individual project. For example, one project may have four buildings—North Building, South Building, East Building, and West Building. Each building is assigned a two-digit number—North Building (01), South Building (02), etc.—which is to follow the two series of numbers previously discussed.

After dividing the project by geographical area, the cost breakdown analysis is further separated into the specific categories of labor hours and the actual cost associated with the amount of time worked on each section of the project for each activity. The breakdown of hours and cost is correlated to the general categories of the productive costs. This organizational format will enable the job to be monitored and updated in accordance with the individual sections and activities associated with the construction project.

Cost Breakdown Assignment Memo

Job No. ____________

	Code	Hours	Dollars
-01—North Building	100-01	2,000	10,000
	200-01	1,000	5,000
	300-01	500	2,500
	400-01	2,000	10,000
	500-01	500	2,500
	600-01	---	---
	700-01	200	1,000
	800-01	---	---
	900-01	---	---
	Total	6,200	31,000
-02—South Building	100-02	4,000	20,000
	200-02	2,000	10,000
	300-02	etc.	etc.
	400-02	.	.
	500-02	.	.
	600-02	.	.
	700-02	.	.
	800-02	.	.
	900-02	.	.
	Total	x,xxx	xx,xxx
-03—East Building	etc.	etc.	etc.
	.	.	.
	.	.	.
	.	.	.
	.	.	.
	.	.	.
	Total	x,xxx	xx,xxx

Your company's controller or finance manager should review and approve the Cost Breakdown Assignment Memo.

The standard job cost code will be utilized for all electrical construction jobs except for deviations granted in writing by the cognizant division controller or the corporate finance manager.

Intermediate breakdown level budgets can be established by use of the Budget Request form.

CONTRACT FUNDING

Purpose

The requirement for the home office and the operating division to be provided with current information (e.g., construction, projects, or operations) regarding contract funding and the status of negotiation of changes, claims, and backcharges is an important function for the funding of the contract.

Policy and Procedure

It is the responsibility of your company's operations or projects group to maintain current contractual funding information organized by individual project and to have overall responsibility for negotiating changes, claims, and backcharges as required.

The contracts administration department or equivalent office should prepare and maintain the necessary files, logs and status reports to provide, in the format presented, monthly information to the managers of contract administration and finance or accounting. An explanation of the Contract Funding Report (Form No. 03) follows:

Changes. This is material and/or work performed or to be performed, authorized or anticipated to be authorized, and said work is outside the scope of the contract.

1. Adjudicated Changes. Funds have been negotiated and written authority has been received to proceed; formal change order has been or will be issued.

2. Pending Negotiation. Funds and/or written authority are still pending negotiation.

3. Probable Changes. The changes log shall include and identify as such all requests for estimates and the estimated

CONTRACT FUNDING REPORT

DIVISION: ______ MONTH ENDING: ______

JOB NO. ______ DESCRIPTION: ______ ORIGNAL CONTRACT AMOUNT:$ ______

A. CHANGES

1. TOTAL SUBMITTED: ______
2. ADJUDICATED: ______
3. PENDING NEGOTIATION: ______
4. ESTIMATED RECOVERY: ______ % OF PENDING NEGOTIATION: ______
5. TOTAL PROBABLE CHANGE (LINE 2 AND LINE 4) ______

B. CLAIMS

6. TOTAL SUBMITTED: ______
7 NEGOTIATED: ______
8. PENDING NEGOTIATION: ______
9. ESTIMATED RECOVERY: ______ % OF PENDING NEGOTIATION: ______
10. TOTAL PROBABLE CLAIMS (LINE 7 AND LINE 9) ______

C. EXTRA WORK

11. BACKCHARGES - INCOME ______ (INCLUDE IN E)
12. BACKCHARGES - OUTGO ______ (DO NOT INCLUDE IN E)

D. TOTAL ADDED CONTRACTUAL INCOME: (LINE 5 PLUS 10 PLUS 13) ______

E. TOTAL PROBABLE CONTRACT AMOUNT: ______

F. OTHER INCOME: ______

G. REMARKS: (LIST PERTINENT CHANGES OR CLAIMS WHICH MAY REQUIRE SPECIAL ATTENTION)

POSSIBLE CHANGES $: ______ POSSIBLE CLAIMS $: ______

______

______

______

Form No. 03: Contract Funding Report

amounts. These dollars will be shown only under "Remarks" on the format attached until such time as positive action is taken on the part of the customer.

Claims. Under most contracts, a claim for an adjustment to the contract price can be developed for work added to or deleted from your contract, defective plans or specifications, changed conditions, untimely delivery of customer furnished equipment, etc.

1. Negotiated Claims. Claim has been honored, dollar amount agreed upon and change order is forthcoming to incorporate it into the contract.
2. Pending Negotiation. Claim has been prepared and submitted but final disposition is still pending.
3. Probable Claim. The claims log shall include and identify as such all areas of probable claims and an estimated claim amount. These dollars will be shown only under "Remarks" on the format attached until such time as claims are prepared and submitted.

NOTE: The subject of claims and their analysis will be discussed in greater detail further in chapters seven and eight of this book.

Backcharges.

1. Income. Work authorized by the customer to be performed by the electrical contractor on the spot, and signed for by the customer as an acceptable charge to the customer.
2. Outgo. Work authorized by the electrical contractor, to be performed by the customer on the spot and signed for by the electrical contractor as an acceptable charge to same. Predetermined rates acceptable to both parties are used to price out such charges.

Total Added Contractual Income. Self-explanatory on format.
Total Probable Contract Amount. Self-explanatory on format.
Other Income. This income has no effect on the contract price. It is a net income figure for

1. Work performed by the corporation for other subcontractors.

2. Work performed for the corporation by other subcontractors.
3. Work performed by the corporation to correct defective or incorrect material, the cost of which will be borne by the supplier.
4. Other.

USE OF THE STANDARD JOB COST CODE FOR TENANT WORK

The standard job cost code provides the cost breakdown structure for use in the establishment of budgets and the accumulation of actual costs for all tenant work jobs.

The Standard Job Cost Code list (below), sets forth the detailed cost account structure which can be used by an electrical contractor to establish budgets and to accumulate direct job costs for all tenant jobs.

The detailed account code structure consists of a two-digit number with a three-digit dash number. The two-digit number is used to segregate productive costs (11 and 21, if needed), subcontract costs (31), and other direct job costs (41). The three-digit dash number further classifies data within these major categories according to individual suites or other detailed job costs.

The standard job cost code should be utilized for all tenant work jobs unless the appropriate manager or controller grants a deviation from this requirement.

Cost breakdown numbers for individual suites are obtained from the cognizant controller or manager through the use of the Budget Request—Tenant Work (Form No. 02, as modified for tenant work) as set forth in the next section.

Tenant Work Standard Job Cost Code

11-000	**PRODUCTIVE COSTS**
-011–700	Individual Suites (Labor, Material and Other Direct Job Costs)
-800	Material Inventory
-900	Craft Supervision
31-000	**SUBCONTRACTS**

Tenant Work Standard Job Cost Code (cont.)

41-000	**OTHER DIRECT JOB COSTS**	
-100	**SUPERVISION AND SUPPORT LABOR**	
	-110	Supervising
	-120	Engineering
	-130	Clerical
	-140	Warehousing
-200	**PAYROLL TAXES, INSURANCE & FRINGE BENEFITS**	
	-210	Payroll Taxes
	-220	Workmens Compensation Insurance
	-230	Public Liability & Property Damage Insurance
	-240	Union Fringe Benefits
	-250	Subsistence and Mileage
-300	**EQUIPMENT**	
	-310	Depreciation and Lease Payment (Includes Rentals)
	-320	Maintenance, Repairs, Fuel and Operation Cost
-400	**SMALL TOOLS AND SUPPLIES**	
-500	**MISCELLANEOUS EXPENSES**	
	-501	Telephone and Telegraph
	-511	Utilities
	-512	Services—Janitorial, Drinking Water, etc.
	-513	Office Supplies and Postage
	-514	Reproduction and Drafting Services
	-515	Office Rentals
	-516	Office Machine Rentals
	-517	Licenses and Permits
	-518	Performance and Bid Bonds
	-519	Legal and Audit
	-520	Insurance Other Than Payroll
	-521	Transportation Costs—Equipment Only
	-522	Outside Engineering and Consulting Services
	-523	Travel and Business Conference Expenses
	-524	Other

Policy Considerations

Job cost control is the principal responsibility of the project manager. Working in this capacity, he oversees the entire scope of contract performance. He prepares tenant work budgets, secures approval, and implements and supervises the performance of the work. He has the responsibility of reviewing all expenditures, evaluating construction progress, and preparing all cost reports and analyses. The timely and efficient performance of this work also involves the establishment of a working relationship with the controller or accounting department in order to fully apprise these groups of any earnings impairment that may occur.

Establishment of Tenant Work Budget

On receipt of a customer request for a price quotation on a suite under a tenant work contract, the project manager computes the standard units and any nonstandard work required, completes a Budget Request—Tenant Work (Form No. 02, as modified for tenant work) and presents a quotation to the customer.

On receipt of a customer notice to proceed, the project manager revises the Budget Request as may be necessary based on the customer notice documents. The project manager then attaches a copy of the customer documents and forwards the Budget Request to the appropriate manager for approval.

This manager approves the budget by affixing his signature and then forwards the budget to the controller or manager of operations or projects. This manager then assigns the suite breakdown number, executes and forwards the original copy to the contracts manager, retains the first copy, and returns the second copy to the project manager.

Development of Original Estimate of Completion

The initial step in the performance of tenant work is the development of a complete cost estimate. This activity involves the following work items.

Initially, the project manager supervises the completion of a material take-off and the calculation of costs for direct labor and material. Based upon the length of time required, the complexity of the job, the number of craft laborers to be employed, and other appropriate factors, the project manager estimates the amount of craft supervision and other direct job costs to be allocated to the tenant work and computes the total estimated direct cost.

The project manager then completes a Monthly Summary of

Job Progress—Tenant Work (Form No. 04) by entering the approved budget in the Budget column under the headings for labor, material and equipment, entering the original estimate at completion in the Estimate at Completion column in the same manner; and computing and entering the variance in the Variance column. The completed report is forwarded to the controller or accounting manager who distributes the Monthly Summary of Job Progress as determined by the manager of projects, construction or operations.

Accounting and Cost Accumulation

The first step in accumulation of tenant work cost occurs when the accounting department charges suite 11-000 breakdown number for the amount of material indicated in the estimate at completion in the initial Monthly Summary of Job Progress.

Daily time sheets are prepared by the job foreman in the prescribed manner charging the time of craft labor to the suite 11-000 breakdown number on which work was performed. Craft supervision is charged to breakdown number 11-900. The project manager should review the daily time sheets for accuracy and completeness and forward them to the accounting department.

The project manager's salary, engineering and other labor, equipment costs, and other miscellaneous direct costs should be charged to the appropriate 41-000 breakdown numbers.

Direct material is procurred in the normal manner and charged to breakdown number 11-800.

The accounting department allocates all costs incurred in cost breakdown numbers 11-900 and 41-000 five days before the end of the accounting month. Costs are allocated to suite breakdown numbers based upon the ratio of craft labor expended on each suite to the total craft labor charged to suites during the month. This allocation results in charges (debits) to each suite account breakdown number and correspondingly equal credit to the 11-900 and 41-000 account breakdowns.

Lagging charges and minor work order costs on completed jobs are charged to cost breakdown 11-799 and combined with completed jobs in the monthly report. Any additional minor work under funding received on completed projects is similarly included with completed projects on the monthly report.

The project manager should verify significant debit balances in breakdown 11-900 monthly by conducting an inventory of material on hand. Variances are charged or credited to the contract cost of sales.

SUBJECT Monthly Summary of Job Progress

(2) JOB NAME ______ (1) REPORT NUMBER ______

(3) CUSTOMER ______ (15) PROGRESS BILLING: ______ (5) WEEK ENDING ______

(4) JOB NUMBER ______ (7) START DATE ______

PREPARED BY ______ (6) CHANGE NUMBER ______ (7) COMPLETION DATE ______

(8)		1 FIRM BUDGET HOURS (9)	2 PRELIM. BUDGET HOURS	3 TOTAL BUDGET HOURS	4 EST. % COMP (10)	5 HOURS EARNED	HOURS PAID						ESTIMATE AT COMPLETION (14)	
							WEEK (11)		INCEPTION TO DATE (12)		ITD (O)/ UNDER HOURS EARNED (13)			
							6 TOTAL	7 OVER-TIME	8 TOTAL	9 OVER TIME	10 HOURS	11 %	12 TOTAL	13 (1)/U BUDGET
10-100	CONDUIT & RACEWAYS													
-200	WIRE & CABLE													
-300	DISTRIBUTION EQUIPMENT													
-400	FINISH WORK													
-500	SPECIAL SERVICE SYSTEMS													
-600	MOTORS & CONTROLS													
-700	GROUNDING & BONDING													
-800	OUTSIDE DISTRIBUTION													
-900	OTHER PRODUCTIVE COSTS													
	TOTAL													

(16) EXPLANATION OF VARIANCE & COMMENTS:

Form No. 04: Monthly Summary of Job Progress—Tenant Work

Reporting Requirements and Methods

The reporting of progress on tenant work should include the same activities as the regular weekly reporting techniques which are discussed in chapter four. The project manager should complete the Weekly Summary of Job Progress—Tenant Work (Form No. 04, modified for tenant work), by the close of business each Friday. This activity should be accomplished every week, as it is very important in the accuracy and reliability of the events of the construction project.

The accounting department should complete the Monthly Summary of Job Progress—Tenant Work (Form No. 04, as modified for tenant work), as of the end of each accounting month. Distribution of the monthly report is made as determined by the division manager.

On completion of work on an individual suite, the project manager should indicate that fact on the Weekly Summary of Job Progress by entering the date complete in the Completion Date column. Completed jobs are deleted from the succeeding weekly report.

If approved supplements are available, as suites are completed and reported in the weekly report the accounting department should render billings to customers. If an approved supplement is not available, the controller should request the contracts manager and the project manager to expedite action on this activity.

On the Monthly Summary of Job Progress form, the months of completion of the completed jobs are reported in the normal manner with the job completion date entered in the appropriate blank. In succeeding months following the month of closure, all completed jobs are consolidated and shown as one entry on the Monthly Summary of Job Progress carrying the description 11-799—Completed Suites.

COST REPORTS AND LABOR PRODUCTIVITY

The Interrelationship of Scheduling and Cost Control

The electrical contractor is often required to perform nonrepetitive work and fatiguing physical labor in an ever-changing construction environment. Since lack of labor productivity and timely recognition of its causes can result in substantial cost overruns, the electrical contractor must establish and enforce accurate and timely cost accounting and project status reporting. It is only through an accurate accounting and cost code procedure compatible with project network scheduling techniques that an electrical contractor can review each project. The contractor must be able to determine if the actual costs are staying within budget and that projected expenditures are based upon the amount of work left to complete as defined by the combination of quantitative measure, unit price history, and most importantly, the time required to complete the activity. In order to monitor labor and other construction costs effectively, the electrical contractor must determine both the costs incurred and committed on a periodic basis and compare them against planned expenditures. The critical nature of this procedure cannot be overemphasized since lack of labor production and timely cost control procedures are key factors in cost overruns and resulting financial difficulties for many electrical contractors.

Unfortunately, many electrical contractors customarily prepare reports on the project schedule status and project cost status separately and do not correlate this information to compare actual construction schedule progress and actual construction cost. A typical project cost report only contains data on the project budget, costs committed to date, estimated cost to complete, estimated final cost, and variances (under and over) between the project budget and the final cost of completion.

This uncoordinated approach to cost and schedule reporting usually results in monthly total cost reports equaling the original budget until the project achieves a 60 to 80 percent completion status. At that stage of the project, there may be a difference between the costs incurred to date and total budgeted cost for a particular activity which result in an overrun.

This historical cost accounting system has two disadvantages. Most importantly, it deprives the project manager of obtaining an accurate picture of project status since the amount of budget earned is never compared to the actual costs committed or expended during the course of construction. Without this information, the project manager cannot evaluate and compare the work accomplished to date against what has been earned from the budget. Therefore, he cannot determine if the actual job progress is within budget or whether the cost of completion will be under or over budget. This procedure also prevents early recognition of inefficiencies and deviations from planned job progress and, thus, makes corrective action more difficult or costly, or both.

The introduction of the earnings factor into the project cost control program provides the extra dimension needed to correlate actual costs expended for a given period of time and the amount earned from the budget for that same period of time. This information provides the data for more accurate forecasting of costs to complete the project, or provides project management with additional information necessary for early recognition of potential cost overruns which may be remedied before they actually occur.

A format for job cost information which incorporates data on actual earnings should include the following categories:

- Description
- Type of Unit (lump sum, square foot, cubic yard, each)
- Bid Quantity
- Estimated Dollars
- Quantity Complete to Date and Percentage
- Amount of Estimate Earned to Date
- Amount Cost to Date
- Amount (Over)/Under Estimate to Date
- Estimated Unit Cost
- Actual Unit Cost to Date
- Unit Cost Variance to Date
- Quantity to Complete
- Projected Unit Cost to Complete
- Project Cost to Complete
- Total Cost at Completion
- (Over)/Under Estimate

A comprehensive cost accounting system for labor productivity also requires a planning and scheduling system which segregates the planned labor hour expenditures on a monthly basis, as well as a cost accounting system to trace planned, earned, and actual expenditures. This procedure allows the project manager to simultaneously analyze actual construction progress and actual cost by using a time-cost relationship commonly referred to as a cumulative "S" curve. This task can be accomplished by extracting the planned man-hours on a weekly or monthly basis from the CPM schedule and plotting them on a cumulative basis over a time graph as shown in Figure 3–1. During actual construction, the project manager can plot actual man-hours expended and the amount of man-hour budget earned for a particular reporting period in order to compare planned, earned, and actual man-hours. This procedure is illustrated in Figure 3–2.

This example allows the contractor to timely identify those performance problems which can potentially result in costly labor overruns. For example, Figure 3–2 presents the following problems:

1. A labor productivity or budget problem exists when comparing the actual man-hour expenditures to the amount of budget earned;

2. The project is behind schedule since the earned man-hours are less than the planned man-hours; and

3. The project cannot be completed according to the original budget.

Factors Which Can Influence Labor Productivity

In order to assess the causes of decreased labor productivity, the electrical contractor must be aware of what factors or job conditions produce or cause an effect on productivity. It should be remembered that these factors or job conditions can be incurred individually or in combination with resultant increases in labor costs.

The electrical contractor must be aware of what factors or other job conditions have historically resulted in loss of labor productivity. These causes often include:

1. *Stacking of Trades.* Additional operations affect the originally planned sequence of work. Contractor may actually have to perform activities concurrently which he originally planned to perform consecutively;

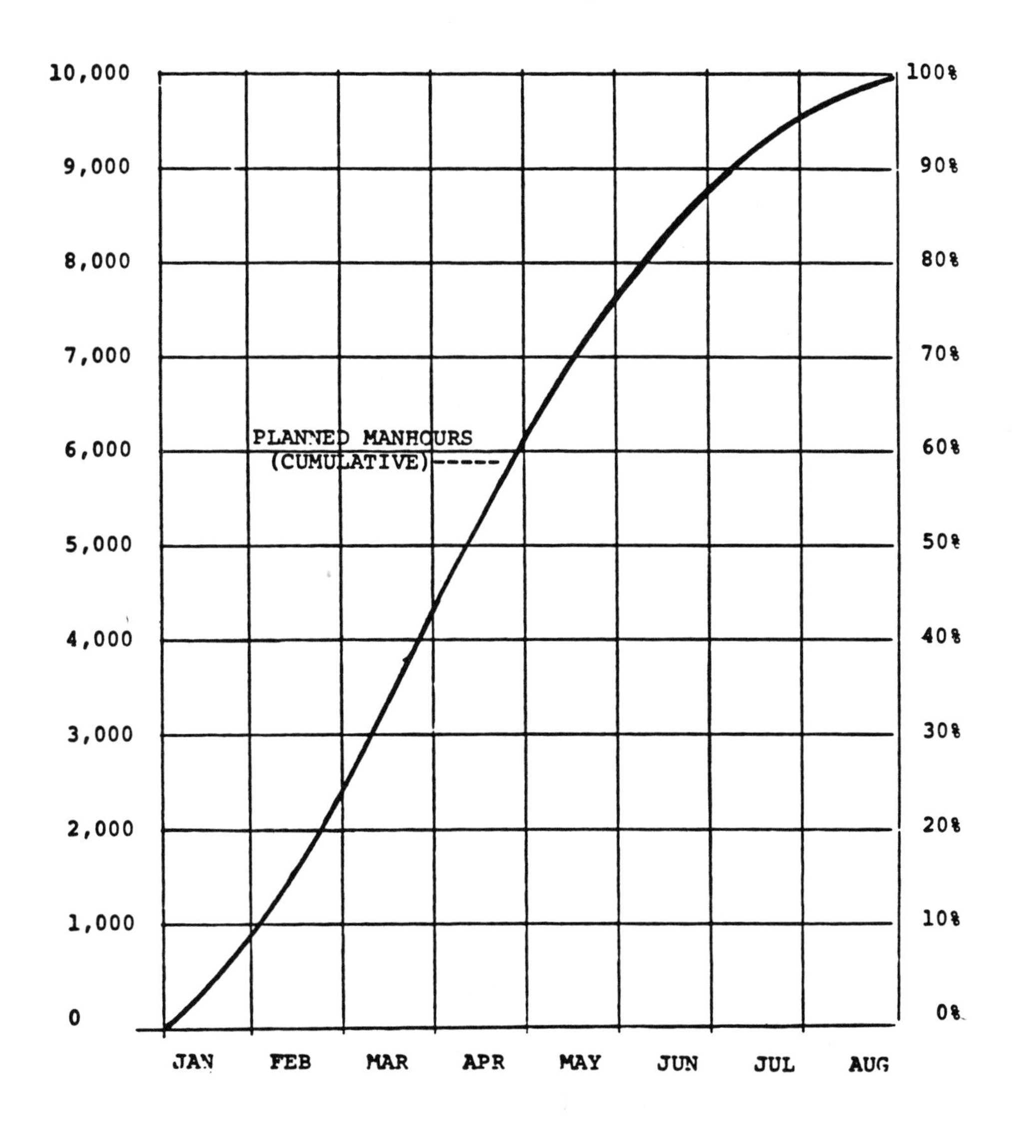

Figure 3–1: Time—Man-Hour Relationship

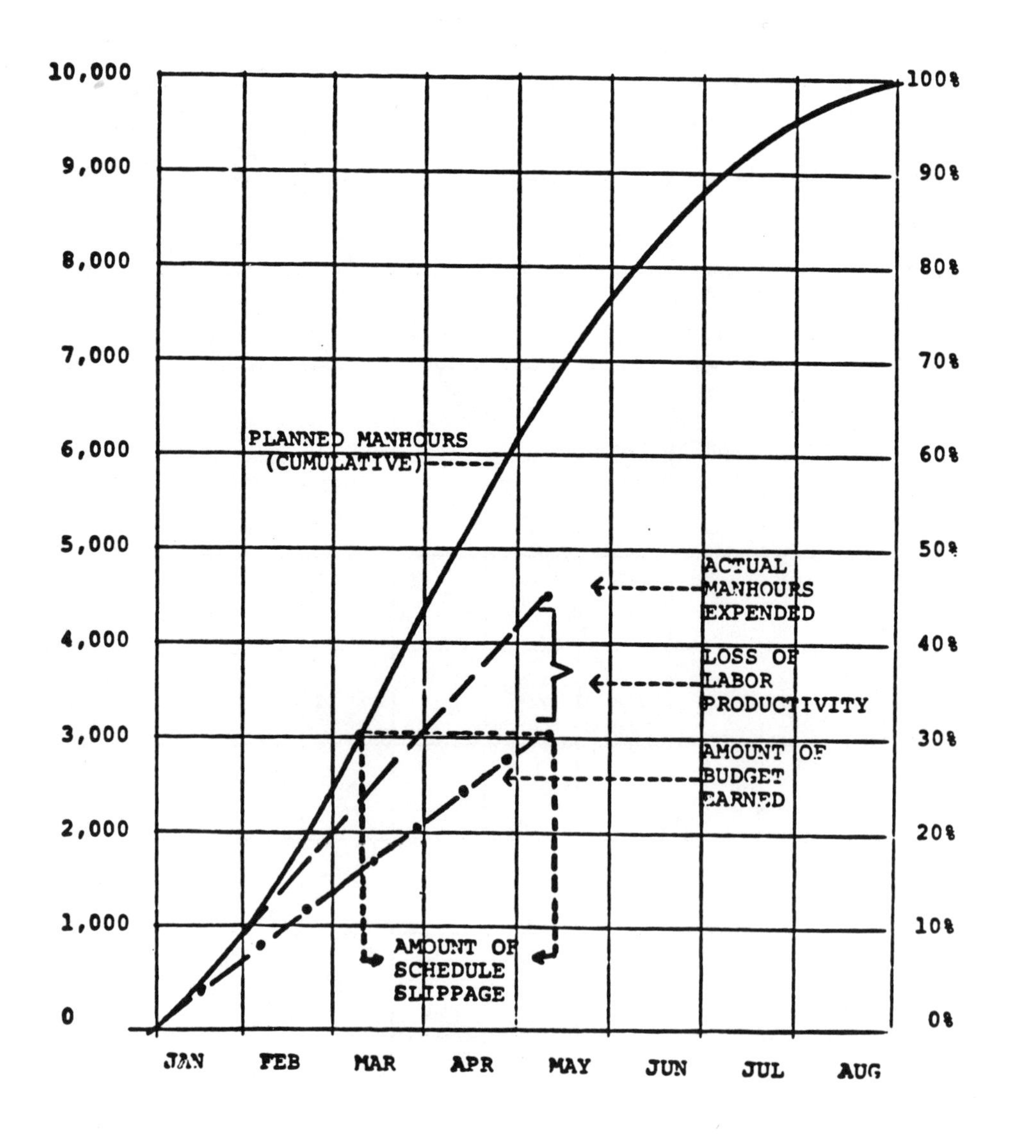

Figure 3–2: Planned, Actual, Earned Man-Hour Relationship

2. *Reassignment of Manpower.* Performance problems such as unanticipated changes, unusual site conditions or lack of coordination may disrupt the planned sequence of electrical construction. This condition may cause additional mobilization and demobilization of manpower on the project;

3. *Limited Site Access.* Conditions on the project such as owner occupancy of the project, lack of material or personnel hoists and stacking of trades potentially reduce efficiency of construction operations;

4. *Fatigue.* The use of the same work forces to perform change order and base contract work may result in physical and mental fatigue;

5. *Overtime.* To maintain production schedules often reduces worker productivity as a result of absenteeism, accidents, job shopping and fatigue;

6. *Larger Crew Size.* The addition of more workmen to existing crews may adversely affect the crew's momentum, mental attitude and overall team effort;

7. *Changes.* Changes to the scope of the work, particularly when coupled with acceleration, may require out-of-sequence performance of the work; and

8. *Weather Change.* Performance of work is scheduled in unusually severe (hot or cold) weather when not originally scheduled to do so.

Project management should be aware of these factors and their possible impact, either individually or in combination, on labor productivity. If the causes for loss of labor productivity are beyond the control or without the fault of the electrical contractor, these conditions may give rise to the entitlement of the electrical contractor to receive additional money and/or time for performance of the work. Of course, the ability of the electrical contractor to recover his increased costs is principally dependent upon his ability to recognize and document these potential labor cost overruns.

Material Control Policy and Procedures

THE IMPORTANCE OF A MATERIAL COMMITMENTS LOG

Each purchasing agent or organization should maintain a Material Commitments Log (Form No. 05) of all items for which a purchase order is issued. A summary of purchasing activity should also be prepared each month, or more frequently if required, and reported to the controller and the accounting department by the second workday of the following accounting month.

The controller and the accounting department should also include the material commitments data in the regular monthly accounting reports.

THE METHODS AND USAGE FOR THE MATERIAL COMMITMENTS LOG

The purchasing agent must also maintain a Material Commitments Log (Form No. 05) for each standard cost code within each job number for which material is being procured.

As purchase orders are issued, the following data is entered in the log:

a. Date of Purchase Order

b. Purchase Order Number

c. Item Number

d. Item Description

e. Quantity

f. Unit Price

g. Total Price

h. Shipping Schedule

MATERIAL COMMITMENTS LOG

JOB NAME: ____________________ JOB NO: ________

CUSTOMER: ____________________ COST CODE: ________

PAGE NO: ________

DATE	P.O. NO.	ITEM NO.	DESCRIPTION	QTY	PRICE		DELIVERY SCHEDULE							
							SHIPMENT		SHIPMENT		SHIPMENT		SHIPMENT	
					UNIT	TOTAL	DATE	AMOUNT	DATE	AMOUNT	DATE	AMOUNT	DATE	AMOUNT

Form No. 05: Material Commitments Log

All change orders must be included on the log to assure the accuracy of the amounts committed.

At the close of business each month, the purchasing agent should:

a. Total the amount committed for each cost breakdown during the month and enter it in the total price column after the last item entry for the month.

b. Add the current month's total to the inception to date total from the prior month and enter the total in the total price column after the current month's total.

c. Copy the Material Commitment Logs using a photo copier and forward the copies to the controller.

The controller should also periodically review the total amount committed on a contract by tabulating the total cost of all purchase orders issued for a particular project and comparing this figure to the total cost indicated on all commitment logs on the contract. Material differences in this information should be resolved by coordination between the controller and the purchasing agent.

The controller also has the responsibility of consolidating the material commitment data into the regular monthly accounting reports.

THE SHOP DRAWING LOG

In order to monitor the processing of shop drawings and product data submittals, the project manager should maintain a Shop Drawing Log (Form No. 05A) which contains the following information: (1) a description of the submittal; (2) the submittal date; (3) the requested return date; (4) the actual return date; (5) the action taken on the submittal; and repetitions of items (2) through (5) for resubmittals.

SHOP DRAWING LOG

JOB NAME : ____________________ JOB NO. : ____________

CUSTOMER : ____________________ PAGE NO. : ____________

SUBMITTAL DESCRIPTION	SUBM. DATE	REQ'D RETURN	ACTUAL RETURN	ACTION TAKEN	RESUBM. DATE	REQ'D RETURN	ACTUAL RETURN	ACTION TAKEN

Form No. 05A: Shop Drawing Log

Daily Labor Distribution Policy and Procedures

THE IMPORTANCE OF DAILY TIME RECORDS

Each foreman must maintain an accurate record of the time expended by each employee under his supervision, prepare a daily time sheet for payroll and labor distribution, and certify the accuracy and validity of the daily time sheet.

The project manager or his designee has the responsibility of auditing the daily time sheets prepared by the foreman, and the project manager must assume full responsibility for their accuracy and validity.

COST CODES FOR ELECTRICAL CONSTRUCTION AND TENANT WORK

Each foreman on a particular project must prepare a Daily Labor Distribution Form in duplicate with the following information. (See Forms No. 06 and No. 06A for examples of these completed forms.)

a. Enter job number.

b. Enter job name.

c. Enter responsible foreman's name.

d. Enter number of sheets used.

e. Enter date.

f. Enter craft such as electrician, carpenter.

g. Enter names of all employees in crew including responsible foreman.

h. Enter job classification such as apprentice (appr); foreman (fore).

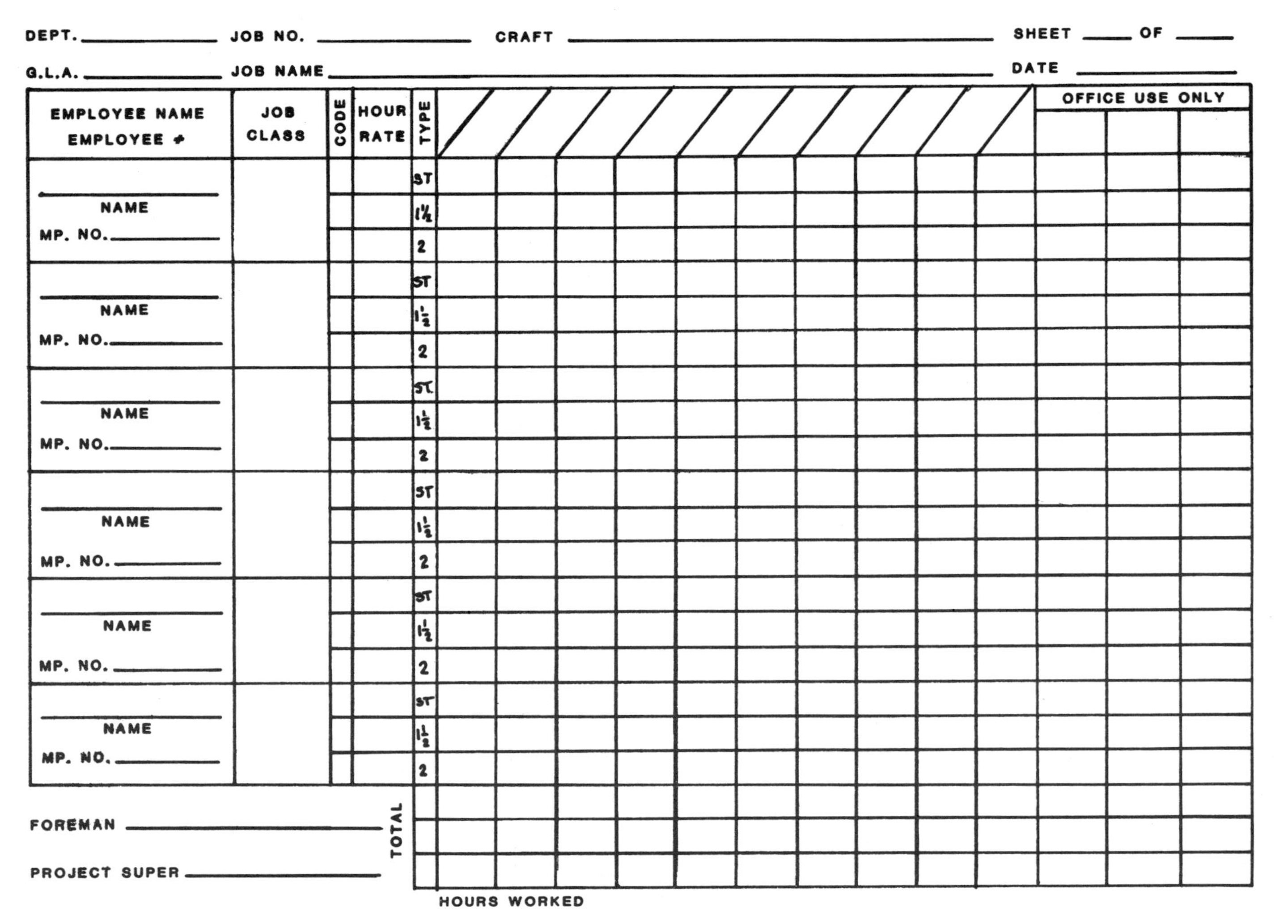

DEPT. ____________ JOB NO. ____________ CRAFT ______________________ SHEET ____ OF ____

G.L.A. ____________ JOB NAME ____ DATE ____________

EMPLOYEE NAME EMPLOYEE #	JOB CLASS	CODE	HOUR RATE	TYPE											OFFICE USE ONLY		
NAME				ST													
MP. NO.				1½													
				2													
NAME				ST													
MP. NO.				1½													
				2													
NAME				ST													
MP. NO.				1½													
				2													
NAME				ST													
MP. NO.				1½													
				2													
NAME				ST													
MP. NO.				1½													
				2													
NAME				ST													
MP. NO.				1½													
				2													
			TOTAL														

HOURS WORKED

FOREMAN ____________________

PROJECT SUPER ____________________

Form No. 06: Daily Labor Distribution Form

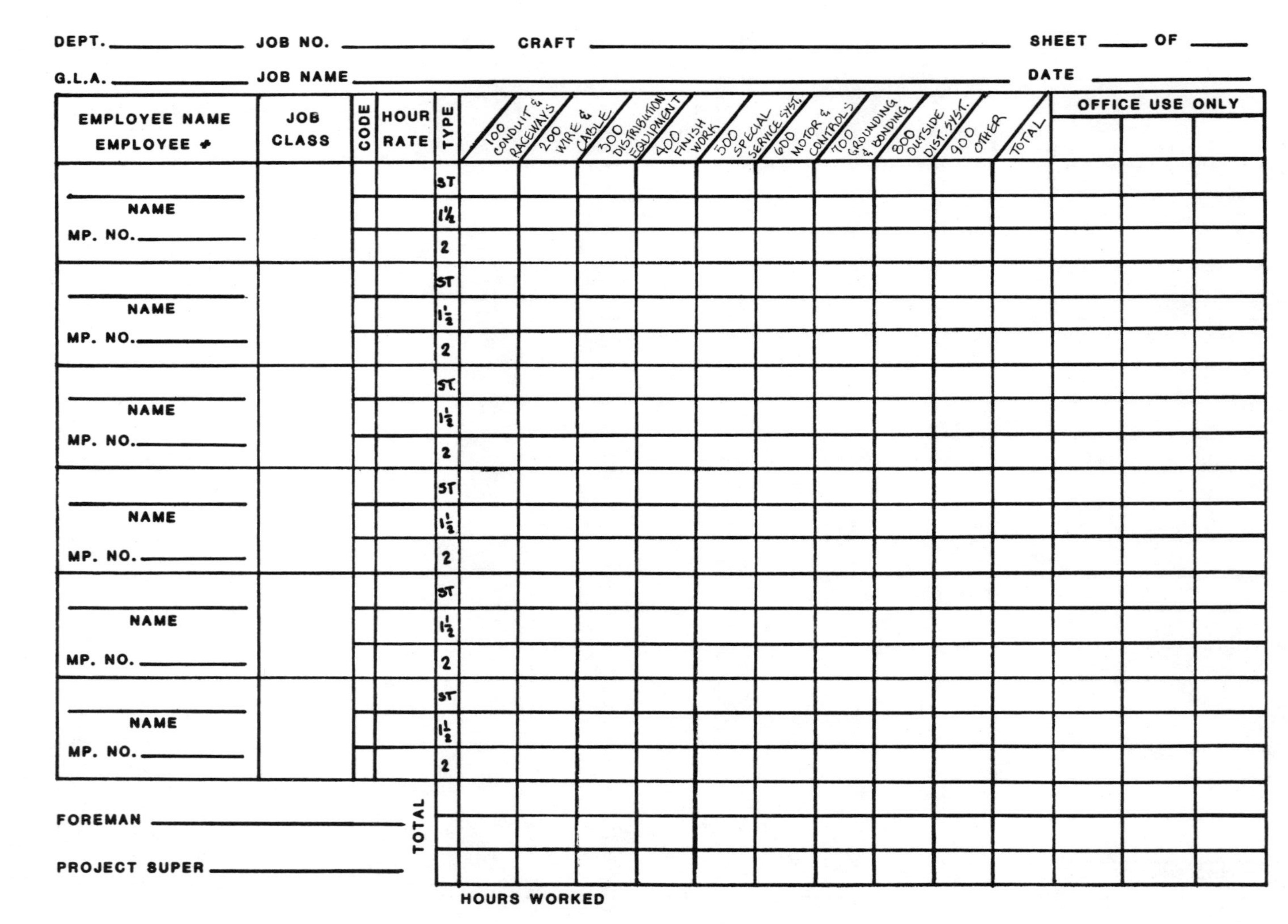

DEPT. ______ JOB NO. ______ CRAFT ______ SHEET ____ OF ____

G.L.A. ______ JOB NAME ______ DATE ______

EMPLOYEE NAME EMPLOYEE #	JOB CLASS	CODE	HOUR RATE	TYPE	100 CONDUIT & RACEWAYS	200 WIRE & CABLE	300 DISTRIBUTION EQUIPMENT	400 FINISH WORK	500 SPECIAL SERVICE SYST.	600 MOTOR & CONTROLS	700 GROUNDING & BONDING	800 OUTSIDE DIST. SYST.	900 OTHER	TOTAL	OFFICE USE ONLY		
NAME				ST													
MP. NO.				1½													
				2													
NAME				ST													
MP. NO.				1½													
				2													
NAME				ST													
MP. NO.				1½													
				2													
NAME				ST													
MP. NO.				1½													
				2													
NAME				ST													
MP. NO.				1½													
				2													
NAME				ST													
MP. NO.				1½													
				2													
			TOTAL														

HOURS WORKED

FOREMAN ______

PROJECT SUPER ______

Form No. 06A: Completed Daily Labor Distribution Form

i. For electrical construction work enter straight time hours worked by each employee at the various cost codes and enter the total in the total column. For tenant work, it is necessary to insert the cost code assigned to the suite in the block at the top of the column under Hours Worked (Exhibit II). Exhibit III sets forth the cost codes to be utilized.

j. Enter hours at premium time rates to be paid employees for overtime or high time.

The foreman should deliver these daily time sheets to the project manager or his designee for approval.

The project manager or his designee also has the responsibility of auditing the time sheets, signifying his approval by signing, forwarding the original document to the appropriate payroll section, and retaining the copy for reference in preparing the Weekly Summary of Job Progress—Man-hours.

Electrical Construction Cost Codes

10-100	Conduit & Raceway
10-200	Wire & Cable
10-300	Distribution Equipment
10-400	Finish Work
10-500	Special Service Systems
10-600	Motors & Controls
10-700	Grounding, Bonding
10-800	Outside Distribution Systems
10-900	Other Productive Costs

Tenant Work Cost Codes

11-011-7000	Individual Suit Numbers as assigned by the Branch Office Manager or Division Controller
11-900	Craft Supervision

Insurance and the Reporting of Damages, Injury, and Losses

Insurance and reports of loss, damage, and injury are necessary in order to achieve proper investigation and salvage disposition; to recover from your insurance carriers; to evaluate the effectiveness of your safety program; and to determine the adequacy of the insurance protection provided.

It is essential that management be informed of every incident that results in damage or injury to the company, other property, or employees, no matter how slight the injury or damage may be. Seemingly insignificant incidents often lead to major claims against the company. Your only reasonable defense of claims is adequate information recorded by eyewitnesses at the time of the incident and promptly reported to the home office.

Usually contained within the general provisions of your subcontract are the AIA General Conditions or similar contract clauses which explicitly state that the general contractor shall have insurance protection in defense of possible claims which could result from the general contractor's operation under the contract. This language is usually incorporated by reference and binding on the electrical subcontractor.

Insurance must encompass the following areas:

1. Workmen's compensation disability benefits, and similar employee benefits acts;

2. Damages due to bodily injury, occupational disease or sickness, or death of employees and claims insured by usual personal injury liability coverage;

3. Bodily injury, sickness or disease, or death of any person other than his employees, and personal liability coverage;

4. Damages due to injury to or destruction of tangible property, including loss of use resulting therefrom.

Due to the extensive insurance coverage and the importance of the coverage, it is essential that any damage or injury be reported immediately. The suggested procedure for reporting the various types of accidents and injuries follows.

REPORTING INJURIES AND ACCIDENTS ON THE PROJECT

Your company's insurance carrier generally furnishes booklets which contain the Supervisor's Report of Injury which is to be completed immediately after each accident involving personal injury on the job. Regardless of severity of injury, each accident must be reported. When medical aid is necessary, a copy of the report, signed by a foreman, must be sent with the injured employee to the doctor contracted for the project. The original of the Supervisor's Report of Injury will be forwarded to the division office for transmittal to the corporate office. The third copy should remain in the booklet for review at the next job site safety meeting. These forms must be complete in every detail, giving all pertinent information available.

Concerning traffic accidents involving bodily injury and/or property damage, the employee is responsible for notifying local police or state highway patrol of any accident involving bodily injury or property damage.

The employee must file the necessary state forms as required under the financial responsibility law of the state where the accident occurs. The required information pertaining to insurance maintained by the corporation is as follows:

Insurance Carrier:	________________
Policy Number:	________________
Policy Term:	________________
Coverage:	Bodily Injury and Property Damage (state the amount required under the financial responsibility law)

A detailed report of *all* accidents must be submitted to the insurance manager in triplicate on Drivers Report of Accident (Form No. 07). If an accident results in serious bodily injury, death, or ex-

DRIVER'S REPORT OF ACCIDENT

YOUR VEHICLE

DATE	TIME A.M. P.M.	LICENCE NO. & STATE
SERIAL NUMBER	COMPANY NO.	MAKE – MODEL
NAME		ADDRESS
SOURCE CODE	DRIVER'S LICENCE	AGE
SPEED MPH	ROAD AND WEATHER CONDITIONS	
DIRECTION OF TRAVEL ( ) NORTH ( ) SOUTH ( ) EAST ( ) WEST		
PROPERTY DAMAGE		
LOCATION OF ACCIDENT (NEAREST STREETS, HIGHWAYS, TOWNS, ETC.)		

OTHER VEHICLE

OWNER'S NAME			ADDRESS
DRIVER'S NAME			ADDRESS
MAKE & MODEL	LICENCE	STATE	SPEED MPH
DIRECTION OF TRAVEL ( )NORTH ( )SOUTH ( )EAST ( )WEST			
PROPERTY DAMAGE			

(IF MORE THAN ONE OTHER CAR IS INVOLVED, USE SPACE FOR OTHER VEHICLE ON AN ADDITIONAL SHEET.)

INJURED PARTIES – NAMES AND ADDRESSES

TAKEN TO – (HOME –HOSPITAL –DOCTOR) AT (LOCATION)

WITNESSES – NAMES AND ADDRESSES

INVESTIGATING OFFICER – NAME & BADGE NO.

DESCRIBE ACCIDENT TO THE BEST OF YOUR KNOWLEDGE – USE OTHER SIDE IF NECESSARY – COMPLETE DIAGRAM ON REVERSE SIDE

DRIVER

Form No. 07: Driver's Report of Accident

tensive property damage, the insurance manager should be notified by telephone immediately as follows:

During business hours: Phone # ________

Other hours: Phone # ________

The employee is to follow up the telephone call with a written report as described earlier. In addition, a copy of the report is filed with the local representative of the insurance company.

On comprehensive insurance loss (i.e., fire, theft, glass breakage and sandstorm), a detailed report is forwarded to the insurance manager on the Insurance Guide (Form No. 08). In case of theft, the local authorities are immediately notified.

For minor damage, such as a broken windshield or a burned cushion, repairs are made and a copy of the paid invoice is forwarded to the insurance manager for reimbursement by the insurance carrier.

For extensive damage caused by fire or water, the employee is to obtain three estimates for the necessary repairs. The estimates are submitted to the insurance manager for authorization prior to repairing the vehicle.

Collision damage to a corporate owned or leased vehicle may be subject to a deductible clause. In those cases where the loss is within the deductible amount, recovery can only be made when liability rests with the other party. When it appears that the other party may be responsible for the accident, estimates should be submitted to the Corporate Insurance Manager as follows:

Two estimates for damages not in excess of $______ .

Three estimates for damages over $______ .

Unless repairs exceed the market value of the vehicle, the repairs should be made by the lowest bidder. Total losses should be reported to the corporate insurance manager.

REPORTING PROPERTY LOSS OR DAMAGE ON THE PROJECT

Your company probably incurs a casualty loss through damage or destruction to property resulting from, but not limited to, such hazards as fire, windstorm, vandalism, malicious mischief, and

INSURANCE GUIDE

REPORTING CASUALTY LOSSES OF PROPERTY

REPLACEMENT COSTS

1. INVENTORIES AND EXPENSED ITEMS

   ACTUAL REPLACEMENT COSTS $ ________________

2. CAPITALIZED ITEMS

   ACTUAL REPLACEMENT COSTS $ ________________

   INSURABLE REPLACEMENT VALUE $ ______________

LABOR

COMPANY LABOR

HOURLY RATE	HOURS	LABOR DOLLARS	OVERHEAD RATE	BURDEN DOLLARS	G & A DOLLARS	NORMAL PROFIT	TOTAL DOLLARS

OUTSIDE LABOR $ ________________________

MATERIAL

COMPANY MATERIAL

TYPE ____

QUANITY ___________________________

COST$ _____________________________

OUTSIDE MATERIAL $ _____________________

DESCRIPTION ____

ADDITIONAL COSTS

ADDITIONAL COSTS $ ______________________________________

DESCRIPTION ____

Form No. 08: Insurance Guide

sprinkler leakage. Casualty losses are reported on property which consists of any tangible item that is owned or leased.

All casualty losses are reported to the insurance manager on the Fire and General Loss Report (see Form No. 09). The following information should be reported on *all* casualty losses.

1. Location
2. Time of loss
3. A description of the loss or damage
4. The estimated cost to repair or replace the casualty loss
5. The person or thing responsible for the loss

The project manager should take prudent action to protect the damaged property and to reduce the loss to a minimum. It should be noted that clean-up labor costs due to damage incurred should be recoverable under your Corporate Insurance Contract.

The authorization necessary to commence with the repairs and/or replacement of damaged and/or lost articles can generally be obtained from your company's insurance manager by the appropriate project manager.

Careful attention must be taken not to destroy any damaged material which should be retained until disposition is authorized by the insurance manager.

As a result of deductible clauses in your company's insurance policies, the amount of the deductible may become a direct job cost.

FIRE AND GENERAL LOSS REPORT

________________ ________
CLAIM NUMBER CODE

CHARGE ALL COSTS TO: ________________ ________________
GENERAL LEDGER ACCOUNT ACCOUNT DETAIL

____
FACILITATION PROJECT NUMBER

TYPE OF LOSS ________________

REPORTED TO CORPORATE INSURANCE OFFICE ________ ____ ____
NAME EXT. DATE

LOSS OCCURED ______ ______ WAS ANYONE INJURED? ________
DATE HOUR

LOCATION ________________

WAS JOB DELAYED OR DISRUPTED? ______ IF SO HOW LONG? ______

ESTIMATED AMMOUNT OF LOST PRODUCTION, IF AVAILABLE ________

DAMAGE TO:

BUILDING	( )	EQUIPMENT	( )	WORK IN PROGRESS	( )
MACHINERY	( )	REPAIR SUPPLIES	( )	PROPERTY OF OTHERS	( )
FUEL	( )				

CAUSE & FACTORS CONTRIBUTING TO EXTENT OF DAMAGE:

____

____

____

____

____

____

LOSS ESTIMATE: $ ____________ $ ____________
PROPERTY DAMAGE BUSINESS INTERRUPTION

____

ADDITIONAL DATA:

DATE: ________

Form No. 09: Fire and General Loss Report

Questions:

1. What is the first step in project cost control?

2. What three categories of work are reflected in the standard cost code system included in this chapter?

3. Who is principally responsible for job cost control on the individual project?

4. What are the two major uses of the cost code breakdown in recording data?

5. How severe must an accident be to merit reporting?

CHAPTER FOUR

Contract Administration

Administration of Contract from Award through Completion

POLICY AND PURPOSE

Contract administration and project controls are required for all work performed solely by your company or work in which your corporation participates (joint venture, etc.). The contract administration or contracts department has primary responsibility for establishing and maintaining such project administration and controls. The contract administration department has the responsibility to assist each project manager in developing procedures to assure performance of the project in accordance with the contract and to provide adequate documentation for the timely execution of changes and claims when there has been deviation from the contract.

RESPONSIBILITIES OF THE CONTRACT ADMINISTRATION DEPARTMENT

The contracts administration, contracts department or its functional equivalent is responsible for the performance of the following activities:

Contract Administration Program Plan

a. At inception of contract, a plan of scheduled events is prepared for the administration of each contract in accordance with the procedures set forth in chapter five. This plan will be updated no less than monthly and submitted to the project manager and director of contract administration or contracts manager.

b. The level of detail utilized in the plan and the contract and number of scheduled events included therein are dependent upon the inclusion of liquidated damages, and the duration, dollar value, and complexity of the contract.

Contract Briefing Paper

A briefing paper is required for each contract. This paper can be written at the bidding stage of the job to assist the estimators and the vice-president of operations or project manager. The document should include at least the following information:

a. Owner, general contractor, contract scope, start and completion date.

b. Unusual specifications, general or special conditions.

c. Other pertinent information concerning payment items, report requirements, etc.

d. Updates as required.

Customer Liaison

a. Introduce key personnel of the electrical contractor and the customer.

b. Establish limits of authority.

c. Originate and answer all correspondence regarding contractual matters.

Administrative Logs and Reports

The following logs will be maintained and, with reports and analyses, will be submitted monthly to the project manager and director of contract administration or contracts manager.

a. Logs to Be Maintained
   1. Correspondence
   2. Change Control
   3. Claims Control
   4. Contracts, Amendments, Deviations or Waivers

b. Reports and Analyses to Be Maintained
   1. Change Status
   2. Claims Status
   3. Cost Status and Analysis—in conjunction with division finance
   4. Job Progress and Analysis—in conjunction with project manager
   5. Backcharges

Other

The contracts administration office will:

a. Visit the job site and review project manager's daily diary as required.

b. Train and assist the project manager in contract matters.

c. Receive from the field office a copy of the following documents:
   1. Weekly Labor Reports
   2. Daily Field Reports
   3. Manloading Charts and revisions thereto
   4. Changes in Prime CPM and Job Progress Performance Schedule as may affect contract conditions
   5. All requests for information

d. Assists in negotiating changes.

e. Prepare and assist in negotiating claims.

f. Provide the accounting department with the necessary information for billing customer when change work has been completed or claims negotiated.

Planning Electrical Construction Projects

REQUIREMENTS OF THE PLAN

A job plan will be prepared for all electrical construction jobs. The minimum detail required for jobs at sales level of $50,000 or less will include calendarized data for total productive labor, total productive material, total subcontract, and total other direct job costs (see Form No. 10). The minimum detail required for jobs over $50,000 is the breakdown outlined in the Budget Request (Form No. 02). Additional detail may be required at the discretion of the project manager.

The following schedule and charts are required. Form numbers indicate samples.

Name	Form Number
Installation Schedule (Bar Chart)	11
Cumulative Physical Percent Complete	12
Completion Status Chart—Cumulative %	13
Monthly Material Committed and Received	14

(1) SUBJECT SUMMARY PLAN - CUMULATIVE

(2) JOB NAME CORAL SANDS HOTEL

(3) CUSTOMER S.D. CLARKE, INC.

(4) JOB NUMBER 442

(5) PREPARED BY WRF (6) APPROVED BY CAS (7) DATE 10/4/77

(8) CHANGE NUMBER 1

(9) START DATE 2/78

(9) COMPLETION DATE 4/79

(15)	(10) Y	ITD	(12) 1978												1979			
	(11) M	—	J	F	M	A	M	J	J	A	S	O	N	D	J	F	M	A
MATERIAL COMM.	P	(13)	460	520	530	540	550	560	565	570	575	580	585	590	593	596	597	601
	A	(14)																
MATERIAL RECEIVED	P	—	10	70	90	140	195	230	265	340	430	455	475	515	563	596	601	603
	A																	
SUB. CONT.	P	—	—	—	—	—	—	—	—	—	—	—	—	—	—	—	—	—
	A																	
LABOR HOURS	P	—	—	0.4	1.8	4.0	7.4	14.8	32.1	33.4	40.6	50.3	67.6	75.1	84.4	96.7	99.4	102.0
	A																	
LABOR $	P	—	—	2.0	9.0	20.0	37.0	74.0	108.9	141.0	301.0	249.5	311.0	376.5	422.1	464.6	497.9	511.1
	A																	
OTHER DIRECT COSTS	P	—	—	2.4	6.1	10.0	14.0	20.2	26.0	31.8	38.5	41.6	51.2	58.5	64.6	69.6	74.8	78.2
	A																	
TOTAL DIRECT COSTS	P	—	10.0	74.4	105.1	170.0	246.0	324.2	399.5	520.8	659.5	749.1	837.2	950.0	1055	1130	1171	1180
	A																	
	P																	
	A																	
PHYSICAL % COMPLETE	P	—	—	—	2	4	7	14	22	29	38	47	61	73	84	89	95	100
	A																	
BILLING %	P	—	—	1	7	10	16	25	33	39	52	64	75	79	88	97	99	100
	A																	
TOTAL	P																	
	A																	

Form No. 10: Summary Plan—Cumulative

Cumulative Material Committed & Received	15
Weekly Manpower Plan	16
Monthly Installation Labor—Hours Paid	17
Cumulative Man-hours Paid	18
Monthly Installation Labor Costs	19
Monthly Other Direct Job Costs	20
Cumulative Other Direct Job Costs	21
Summary Plan—Cumulative	10
Daily Job Diary	22
Weekly Summary of Job Progress—Man-hours	4
Other charts and schedules as required to provide adequate visibility of costs and completion status necessary to proper management and cost control for the job.	

The required planning forms based upon all available data are completed by the project manager and field superintendent prior to the physical move onto the job. In the event that customer data is insufficient for a reasonable planning effort, the project manager may request a 30 day extension from higher management to perform this activity. In any event the planning forms will be completed no later than 30 days after the commencement of construction.

The project manager or his designee should revise the planning charts within 10 working days after any of the following circumstances or events occur:

1. A significant change occurs in the scope of work as indicated by an approved revision to the job budget supported by an approved Budget Request (Form No. 02).

2. Actual incurred costs or physical completion status varies significantly from the plan so that it does not represent an accurate projection of future costs and events.

3. The division manager or the division controller requests revisions.

4. It becomes evident that the job cannot be accomplished within the approved budget. Since planned expenditures should never exceed the authorized budget as represented by a summation of the approved Budget Request, a new Budget Request is prepared and forwarded to the division manager for the necessary approval as described in chapter two.

IMPLEMENTING AND UPDATING THE PLAN

The Use of Standard Forms

Immediately upon receipt of a new job, the assigned project manager and job superintendent, together with the responsible estimator, develop the data necessary to complete the required planning forms. The planning forms are completed according to the instructions in the accompanying exhibits and are presented to the division or projects manager for approval.

The division or projects manager reviews the plan and approves it only when he is satisfied that it properly represents a reasonable and accurate presentation of the expenditures, completion rate, and billing schedule for the job.

The approved plan is forwarded to the area office manager who distributes it to the accounting department.

Revisions to the plan which are made as previously described require the same procedures to be followed.

INSTALLATION SCHEDULE

FORM NO. 11

The explanations below are keyed to the numbers encircled on the face of the Installation Schedule chart. The form used for this chart is a standard form indicating the required breakdown for electrical construction jobs. The form will accommodate a job running for 18 months. If a longer job is being planned, a second page will be necessary.

(1) Since this is a multipurpose form, it is necessary to write in the subject of the chart, which is "Installation Schedule."

(2) Indicate the job name as set forth on the Budget Request Form.

(3) Indicate the name of the owner for whom the work is being performed.

(4) Indicate the job number as set forth on the approved Budget Request.

(5) Indicate your initials as the preparer.

(6) Indicates approval of the branch manager when properly initialed.

(7) Indicate date the schedule is prepared.

(8) Include the change number per the approved Budget Request form.

(9) Indicate start and completion dates.

(10) Insert proper years and a vertical line to separate the years at the appropriate places.

(11) Signify months by appropriate abbreviation.

(12) Inception to date data goes in this column. It is not used for this schedule.

(13) Draw a bar in the horizontal row P representing the time span over which the work will be accomplished.

(1) SUBJECT INSTALLATION SCHEDULE

(2) JOB NAME CORAL SANDS HOTEL

(3) CUSTOMER S.D. CLARKE, INC.

(4) JOB NUMBER 442

(5) PREPARED BY WRF (6) APPROVED BY CAS (7) DATE 10/4/77

(8) CHANGE NUMBER 1

(9) START DATE 2/78

(9) COMPLETION DATE 4/79

(15)		(10) Y / (11) M	ITD	(12) 1978 J	F	M	A	M	J	J	A	S	O	N	D	1979 J	F	M	A
10–100	CONDUIT & RACEWAY	P	(13)																
		A	(14)																
-200	WIRE & CABLE	P																	
		A																	
-300	DISTRIBUTION EQUIPMENT	P																	
		A																	
-400	FINISH WORK	P																	
		A																	
-500	SPECIAL SERVICE SYST.	P																	
		A																	
-600	MOTOR & CONTROLS	P																	
		A																	
-700	GROUNDING & BONDING	P																	
		A																	
-800	OUTSIDE DIST. SYST.	P																	
		A																	
-900	OTHER	P																	
		A																	
		P																	
		A																	
TOTAL		P																	
		A																	

Form No. 11: Installation Schedule

(14) The horizontal row A is for use in recording actual performance.

(15) This column contains the descriptions of the standard breakdowns that will be planned. Detailed descriptions of these breakdowns are contained in chapter three.

CUMULATIVE PHYSICAL PERCENT COMPLETE

FORM NO. 12

Explanations are keyed in the same manner described for Installation Schedule, Form No. 11.

(1) Enter "Cumulative Physical Percent Complete."

(2) through
(11) Same as Installation Schedule, Form No. 11.

(12) This column is for inception to date data. If this is a new plan, a dash should be entered to indicate that it is not applicable. If it is a revised plan, the cumulative inception to date physical completion percentage will be entered in this column opposite the letter P.

(13) Enter the percentage that will be physically completed at the end of each month under the appropriate month. The last month should equal 100 percent.

(14) through
(15) Same as Installation Schedule, Form No. 11.

(16) In a blank space in the description column enter the heading "Billing Percent", and opposite the P enter the percent that will be billed at the end of each month. If this is a revised plan, the amount billed inception to date is entered in the ITD column, and the remainder is indicated in the appropriate months to total 100 percent billing.

(1) SUBJECT CUMULATIVE PHYSICAL % COMPLETE

(2) JOB NAME CORAL SANDS HOTEL

(3) CUSTOMER S.D. CLARKE, INC.

(4) JOB NUMBER 442

(5) PREPARED BY WRF (6) APPROVED BY CAS (7) DATE 10/4/77

(8) CHANGE NUMBER 1

(9) START DATE 2/78

(9) COMPLETION DATE 4/79

(15)	(10) Y	ITD	(12) 1978												1979			
	(11) M	—	J	F	M	A	M	J	J	A	S	O	N	D	J	F	M	A
10–100 CONDUIT & RACEWAY	P	(13)	—	3	15	20	25	50	65	80	90	94	97	100	—	—	—	—
	A	(14)																
-200 WIRE & CABLE	P	—	—	—	—	5	17	30	41	50	64	74	79	86	91	96	100	—
	A																	
-300 DISTRIBUTION EQUIPMENT	P	—	—	—	4	16	30	45	62	74	87	91	96	99	100	—	—	—
	A																	
-400 FINISH WORK	P	—	—	—	—	—	—	—	—	—	3	10	21	34	62	84	98	100
	A																	
-500 SPECIAL SERVICE SYST.	P	—	—	—	—	—	8	16	35	48	59	72	83	90	95	100	—	—
	A																	
-600 MOTOR & CONTROLS	P	—	—	—	—	—	—	2	10	26	52	78	84	97	100	—	—	—
	A																	
-700 GROUNDING & BONDING	P	—	—	—	—	—	4	14	30	44	62	77	88	90	94	96	100	—
	A																	
-800 OUTSIDE DIST. SYST.	P	—	—	—	—	—	—	—	—	5	25	50	75	90	93	100	—	—
	A																	
-900 OTHER	P	—	—	10	21	31	42	52	61	67	71	78	84	87	92	95	99	100
	A																	
TOTAL	P	—	—	—	2	4	7	14	21	27	39	48	60	72	89	95	99	100
	A																	
BILLING % (16)	P	—	—	1	7	10	23	37	51	62	72	79	84	94	97	98	99	100
	A																	

Form No. 12: Cumulative Physical Percent Complete

COMPLETION STATUS CHART—CUMULATIVE PERCENT COMPLETE

FORM NO. 13

The explanations outlined below are keyed to the encircled numbers on the face of the Completion Status Chart—Cumulative Percent Complete. The form used for this chart is a standard multipurpose form. The form will accommodate a job running for 18 months. If a longer job is being planned, a second page will be required.

(1) through
(9) Same as Installation Schedule, Form No. 11.

(10) This is the legend block. Enter data as indicated on the face of the sample chart:
Dashed line indicates physical completion—Plan;
Solid line indicates physical completion—Actual;
Dashed line with X indicates billing—Plan;
Solid line with X indicates billing—Actual.
The same headings are indicated in the lower left-hand corner, and numerical values of the cumulative percentages are entered prior to posting the lines on the chart. These are based on data developed in the Cumulative Physical Percent Complete schedule (Form No. 12).

(11) Enter month and year designation, separating the years at the appropriate place with a vertical line.

(12) Plot plan lines using appropriate coded lines.

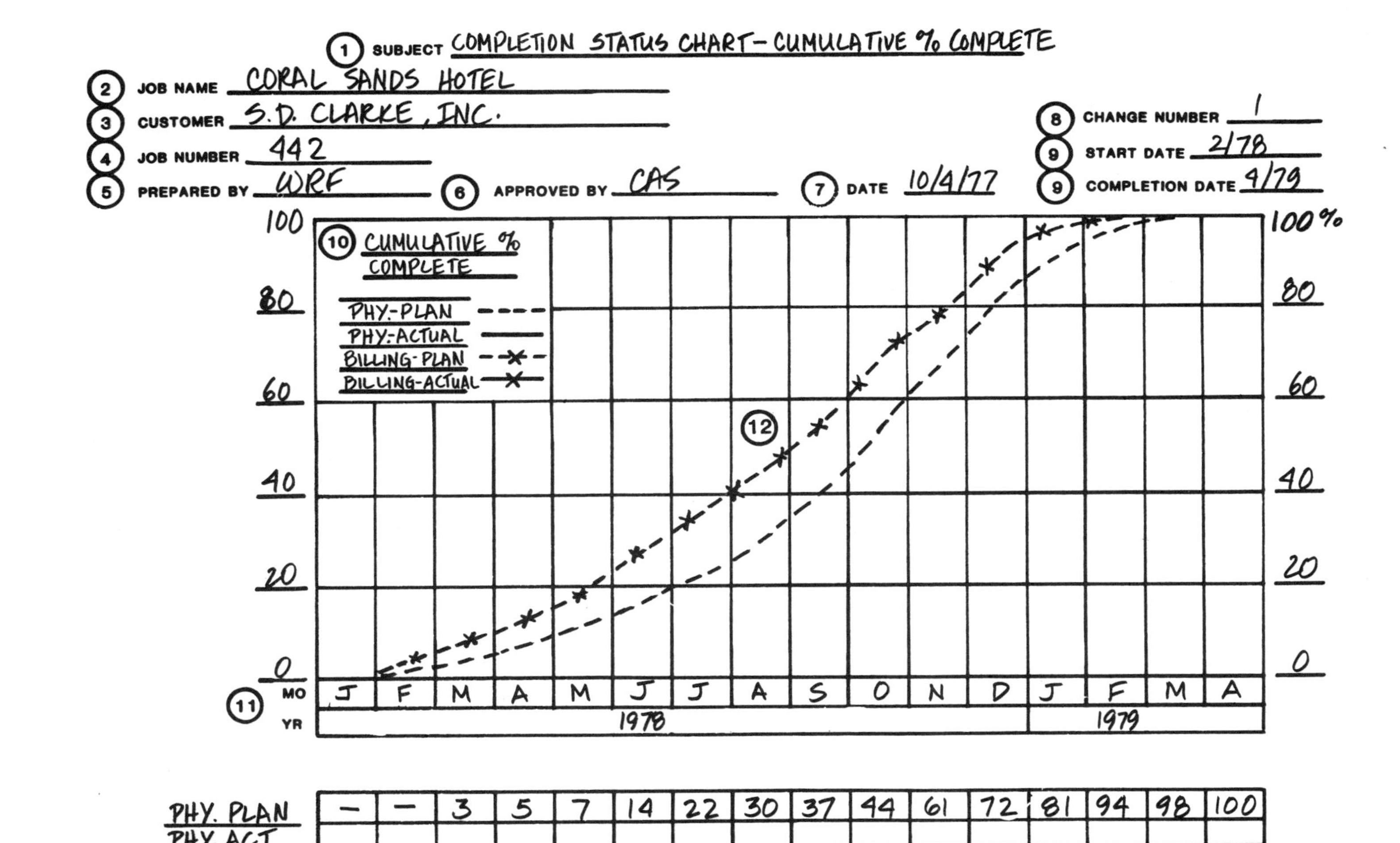

(1) SUBJECT COMPLETION STATUS CHART—CUMULATIVE % COMPLETE

(2) JOB NAME CORAL SANDS HOTEL

(3) CUSTOMER S.D. CLARKE, INC.

(4) JOB NUMBER 442

(5) PREPARED BY WRF

(6) APPROVED BY CAS

(7) DATE 10/4/77

(8) CHANGE NUMBER 1

(9) START DATE 2/78

(9) COMPLETION DATE 4/79

(10)	J	F	M	A	M	J	J	A	S	O	N	D	J	F	M	A
PHY. PLAN	—	—	3	5	7	14	22	30	37	44	61	72	81	94	98	100
PHY. ACT																
BILL-PLAN	—	1	6	11	16	25	34	47	51	62	72	88	94	98	99	100
BILL-ACT																

Form No. 13: Completion Status Chart—Cumulative Percent Complete

MONTHLY MATERIAL COMMITTED AND RECEIVED
FORM NO. 14

The explanations below are keyed in the same manner as for Installation Schedule, Form No. 11.

(1) Enter title "Monthly Material Committed and Received."

(2) through
(11) Same as Installation Schedule, Form No. 11.

(12) Inception to data is entered in this column. There will be no entry on a new job. On revisions, enter the amount committed to date.

(13) Enter the amount to be committed each month. This will normally be rounded to the nearest hundred, or thousand, depending upon the size of the contract. An appropriate indication should be made under the title block.

(14) For reporting actual costs per chapter three.

(15) Write in "committed." This is the standard cost breakdown for electrical construction jobs.

(16) Write in "received" in a blank in the description column. Enter on the plan line the total dollar amount of material you plan to receive under the appropriate month.

(1) SUBJECT MONTHLY MATERIAL COMMITTED & RECEIVED

(2) JOB NAME CORAL SANDS HOTEL

(3) CUSTOMER S.D. CLARKE, INC.

(4) JOB NUMBER 442

(5) PREPARED BY WRF (6) APPROVED BY CAS (7) DATE 10/4/77

(8) CHANGE NUMBER 1

(9) START DATE 2/78

(9) COMPLETION DATE 4/79

COMM. (15)	(10) Y	ITD	(12) 1978												1979			
	(11) M	—	J	F	M	A	M	J	J	A	S	O	N	D	J	F	M	A
10–100 CONDUIT & RACEWAY	P	(13)	50	—	—	—	—	—	—	—	—	—	—	—	—	—	—	—
	A	(14)																
-200 WIRE & CABLE	P	—	40	—	—	—	—	—	—	—	—	—	—	—	—	—	—	—
	A																	
-300 DISTRIBUTION EQUIPMENT	P	—	30	—	—	—	—	—	—	—	—	—	—	—	—	—	—	—
	A																	
-400 FINISH WORK	P	—	70	—	—	—	—	—	—	—	—	—	—	—	—	—	—	—
	A																	
-500 SPECIAL SERVICE SYST.	P	—	100	—	—	—	—	—	—	—	—	—	—	—	—	—	—	—
	A																	
-600 MOTOR & CONTROLS	P	—	50	—	—	—	—	—	—	—	—	—	—	—	—	—	—	—
	A																	
-700 GROUNDING & BONDING	P	—	30	—	—	—	—	—	—	—	—	—	—	—	—	—	—	—
	A																	
-800 OUTSIDE DIST. SYST.	P	—	80	—	—	—	—	—	—	—	—	—	—	—	—	—	—	—
	A																	
-900 OTHER	P	—	10	10	10	10	10	10	5	5	5	5	5	3	3	3	2	2
	A																	
TOTAL COMM.	P	—	460	60	10	10	10	10	5	5	5	5	5	3	3	3	2	2
	A																	
RECEIVED (16)	P	—	10	60	20	50	55	35	35	20	40	53	43	28	3	3	2	2
	A																	

Form No. 14: Monthly Material Committed and Received

CUMULATIVE MATERIAL COMMITTED AND RECEIVED

FORM NO. 15

The explanations below are keyed in the same manner as for Installation Schedule, Form No. 11.

(1) Enter the title "Cumulative Material Committed and Received."

(2) through
(9) Same as Installation Schedule, Form No. 11.

(10) This is the legend block—Enter data as indicated on face of sample chart:

Dashed line indicates cumulative commitments—Plan;
Solid line indicates cumulative commitments—Actual;
Dashed line with X indicates cumulative material received—Plan;
Solid line with X—cumulative material received—Actual.

Enter the same headings at the lower left and enter cumulative figures under the appropriate months based on data developed in the Monthly Material Committed & Received Schedule (Form No. 14).

(11) Enter months and years, separating years by vertical lines at appropriate places.

(12) Plot plan lines using appropriate coded lines.

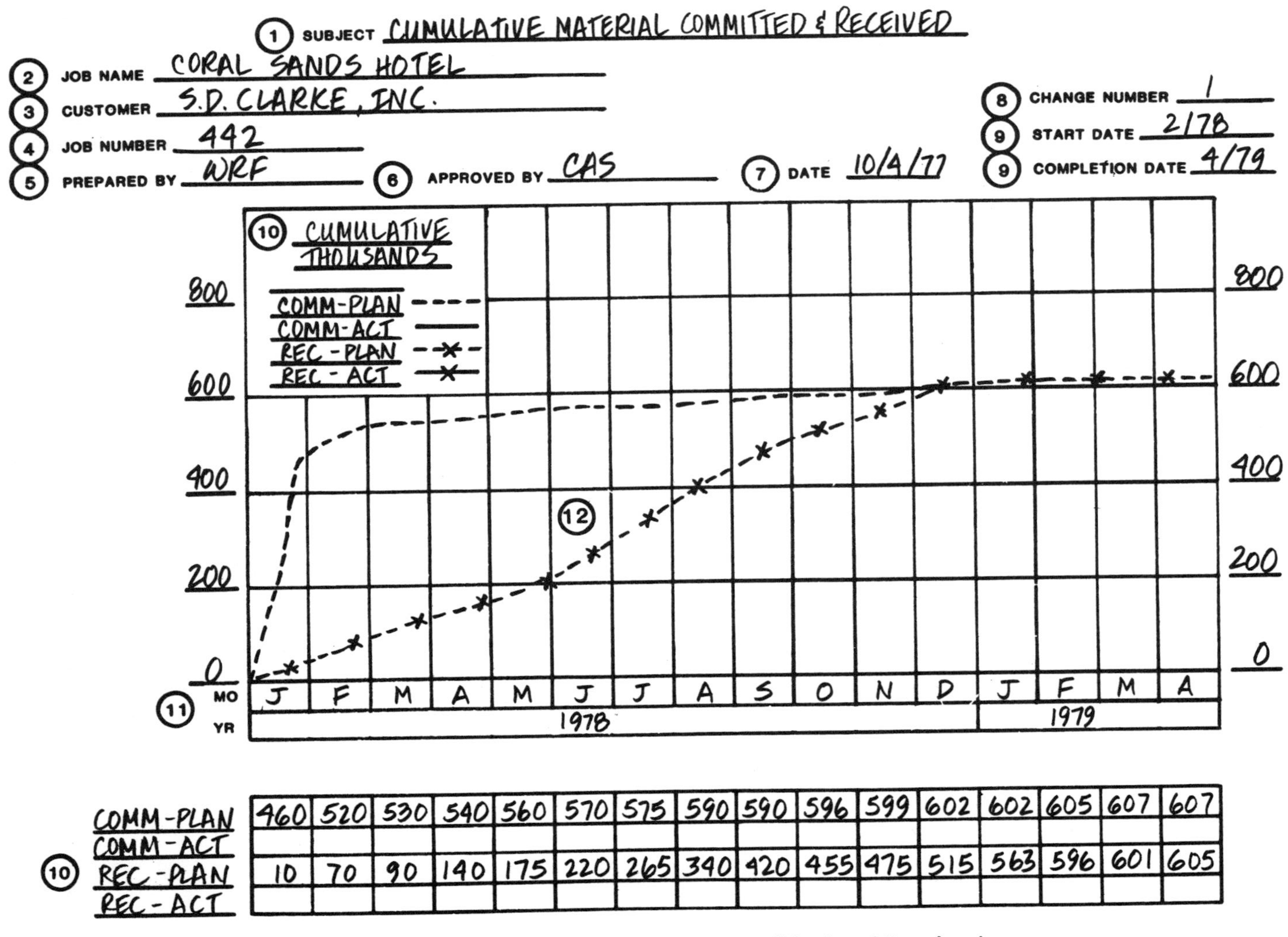

(1) SUBJECT CUMULATIVE MATERIAL COMMITTED & RECEIVED

(2) JOB NAME CORAL SANDS HOTEL

(3) CUSTOMER S.D. CLARKE, INC.

(4) JOB NUMBER 442

(5) PREPARED BY WRF

(6) APPROVED BY CAS

(7) DATE 10/4/77

(8) CHANGE NUMBER 1

(9) START DATE 2/78

(9) COMPLETION DATE 4/79

(10)	J	F	M	A	M	J	J	A	S	O	N	D	J	F	M	A
COMM-PLAN	460	520	530	540	560	570	575	590	590	596	599	602	602	605	607	607
COMM-ACT																
REC-PLAN	10	70	90	140	175	220	265	340	420	455	475	515	563	596	601	605
REC-ACT																

Form No. 15: Cumulative Material Committed and Received

WEEKLY MANPOWER PLAN
FORM NO. 16

The explanations are keyed in the same manner as described for Installation Schedule, Form No. 11. This form will accommodate a four-month period. For longer jobs, additional forms will be required. It should be noted that the chart is established to accomodate a five-week month in the fourth month block.

(1) Enter page number of the total weekly manpower schedule.

(2) through
(9) Same as Installation Schedule, Form No. 11.

(10) Indicate month and year. The last month of each calendar quarter is a five-week month according to the accounting calendar (March, June, September, and December). Data should be entered so that these months fall in the third and sixth month blocks.

(11) Indicates the week end date.

(12) Enter the number of men planned for each week.

(13) For use in indicating actual.

(14) Standard cost breakdown.

(1) SUBJECT WEEKLY MANPOWER PLAN

(2) JOB NAME CORAL SANDS HOTEL

(3) CUSTOMER S.D. CLARKE, INC.

(4) JOB NUMBER 442

(5) PREPARED BY WRF (6) APPROVED BY CAS (7) DATE 10/4/77

(8) CHANGE NUMBER 1

(9) START DATE 2/78

(9) COMPLETION DATE 4/79

(14)		(10) Y	ITD	(12) 1/78			2/78				3/78				4/78				
		(11) M	1/7	1/14	1/21	1/28	2/4	2/11	2/18	2/25	3/4	3/11	3/18	3/25	4/1	4/8	4/15	4/22	4/29
10-100	CONDUIT & RACEWAY	P	(12)	—	—	—	—	—	—	5	5	5	5	5	5	5	5	5	5
		A	(13)																
-200	WIRE & CABLE	P	—	—	—	—	—	—	—	—	—	—	—	—	—	5	5	5	5
		A																	
-300	DISTRIBUTION EQUIPMENT	P	—	—	—	—	—	—	—	—	—	2	2	3	5	5	5	5	5
		A																	
-400	FINISH WORK	P	—	—	—	—	—	—	—	—	—	—	—	—	—	—	—	—	—
		A																	
-500	SPECIAL SERVICE SYST.	P	—	—	—	—	—	—	—	—	—	—	—	—	5	5	5	10	10
		A																	
-600	MOTOR & CONTROLS	P	—	—	—	—	—	—	—	—	—	—	—	—	—	—	—	—	5
		A																	
-700	GROUNDING & BONDING	P	—	—	—	—	—	—	—	—	—	—	—	—	5	5	5	5	5
		A																	
-800	OUTSIDE DIST. SYST.	P	—	—	—	—	—	—	—	—	—	—	—	—	—	—	—	—	—
		A																	
-900	OTHER	P	—	—	—	—	1	1	1	1	1	1	1	1	1	1	1	1	1
		A																	
		P																	
		A																	
TOTAL		P	—	—	—	—	1	1	1	6	6	8	8	9	21	26	26	31	36
		A																	

Form No. 16: Weekly Manpower Plan

MONTHLY INSTALLATION LABOR—HOURS PAID

FORM NO. 17

The explanations below are keyed in the same manner as for Installation Schedule, Form No. 11.

(1) Enter title "Monthly Installation Labor Hours Paid."

(2) through
(11) Same as for Installation Schedule, Form No. 11.

(12) Inception to date data, not used for new job. If revised plan, enter total labor hours paid inception to date opposite P.

(13) Enter the planned hourly expenditures including overtime for each month in the appropriate box opposite P.

(14) Used for recording actual expenditure.

(15) Standard cost breakdown.

(1) SUBJECT MONTHLY INSTALLATION LABOR - HRS. PD.

(2) JOB NAME CORAL SANDS HOTEL

(3) CUSTOMER S.D. CLARKE, INC.

(4) JOB NUMBER 442

(5) PREPARED BY WRF (6) APPROVED BY CAS (7) DATE 10/4/77

(8) CHANGE NUMBER 1

(9) START DATE 2/78

(9) COMPLETION DATE 4/79

(15)	(10) Y	ITD	(12) 1978												1979			
	(11) M	—	J	F	M	A	M	J	J	A	S	O	N	D	J	F	M	A
10-100 CONDUIT & RACEWAY	P	(13)	—	5	20	25	20	40	35	35	20	10	10	5	—	—	—	—
	A	(14)																
-200 WIRE & CABLE	P	—	—	—	—	20	20	20	20	20	20	20	10	10	10	10	8	—
	A																	
-300 DISTRIBUTION EQUIPMENT	P	—	—	—	7	25	20	25	20	20	10	8	5	5	5	—	—	—
	A																	
-400 FINISH WORK	P	—	—	—	—	—	—	—	—	—	10	20	40	40	40	40	40	40
	A																	
-500 SPECIAL SERVICE SYST.	P	—	—	—	—	35	20	40	45	45	90	70	70	90	45	45	70	25
	A																	
-600 MOTOR & CONTROLS	P	—	—	—	—	5	—	7	20	40	40	40	40	20	20	8	5	—
	A																	
-700 GROUNDING & BONDING	P	—	—	—	—	25	20	20	20	20	30	30	30	20	20	20	10	—
	A																	
-800 OUTSIDE DIST. SYST.	P	—	—	—	—	—	—	—	—	—	—	20	70	90	70	70	20	10
	A																	
-900 OTHER	P	—	—	4	4	5	5	5	5	5	5	5	5	5	—	5	—	5
	A																	
	P																	
	A																	
TOTAL	P	—	—	9	31	140	105	157	165	185	225	223	280	285	210	198	153	80
	A																	

Form No. 17: Monthly Installation Labor—Hours Paid

CUMULATIVE MAN-HOURS PAID

FORM NO. 18

The explanations below are keyed in the same manner as for the Installation Schedule, Form No. 11.

(1) Enter the title "Cumulative Man-Hours Paid."

(2) through
(9) Same as Installation Schedule, Form No. 11.

(10) Legend block. Enter data as indicated on the face of the sample chart:
Dashed line indicates planned hours;
Solid line indicates actual incurred hours;
Dashed line with X indicates hours paid including overtime;
Solid line with X indicates hours earned (computed by multiplying percent complete times budgeted hours).
The same headings are entered at the lower left, and the planned cumulative hours are entered opposite Plan under the appropriate month based on data developed in the Monthly Installation Labor Hours Paid Schedule (chapter three).

(11) Enter months and years, separating years at the appropriate place with a vertical bar.

(12) Plot plan line using appropriate coded line.

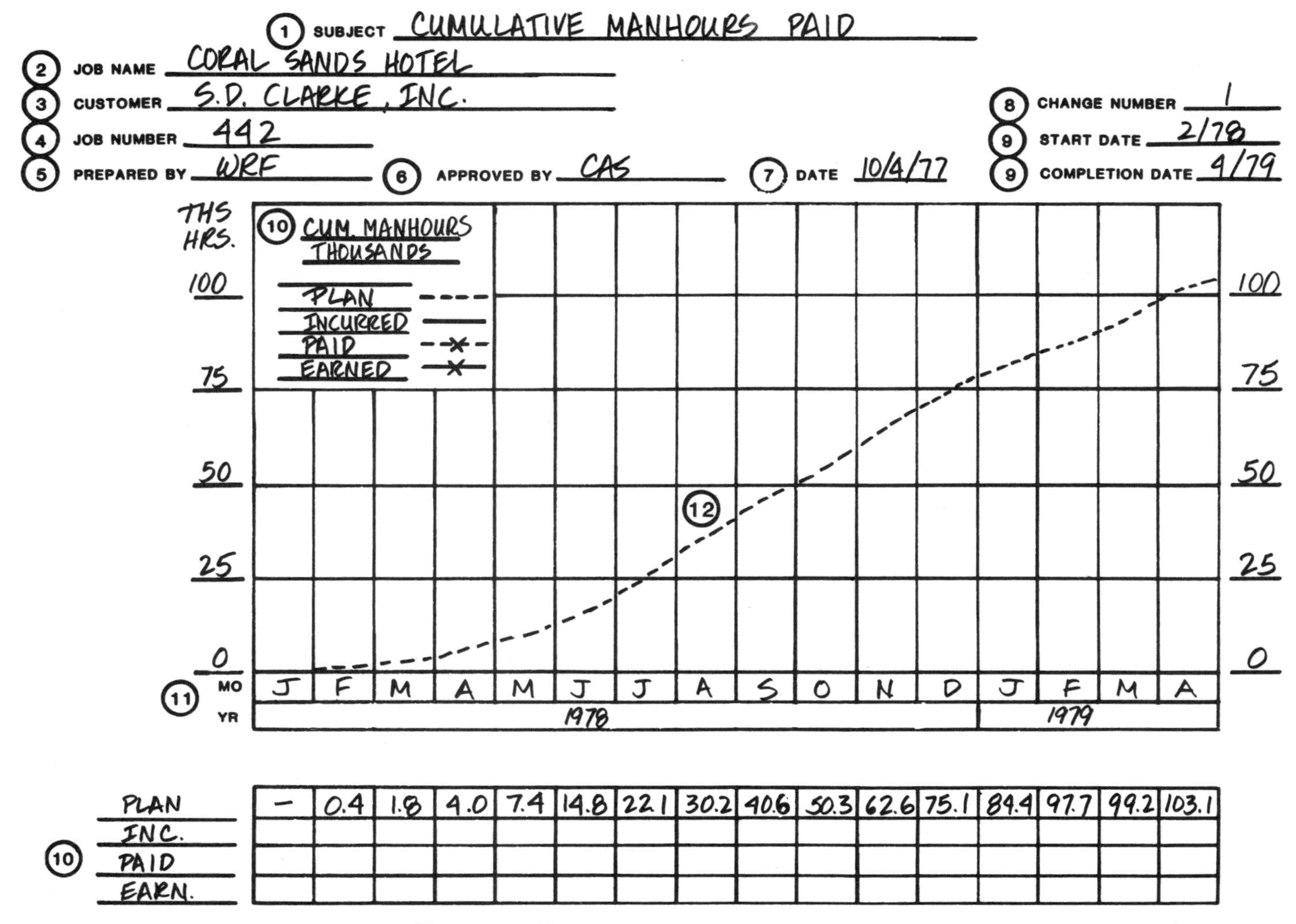

① SUBJECT CUMULATIVE MANHOURS PAID

② JOB NAME CORAL SANDS HOTEL
③ CUSTOMER S.D. CLARKE, INC.
④ JOB NUMBER 442
⑤ PREPARED BY WRF ⑥ APPROVED BY CAS ⑦ DATE 10/4/77

⑧ CHANGE NUMBER 1
⑨ START DATE 2/78
⑨ COMPLETION DATE 4/79

⑩	J	F	M	A	M	J	J	A	S	O	N	D	J	F	M	A
PLAN	–	0.4	1.8	4.0	7.4	14.8	22.1	30.2	40.6	50.3	62.6	75.1	84.4	97.7	99.2	103.1
INC.																
PAID																
EARN.																

Form No. 18: Cumulative Man-Hours Paid

MONTHLY INSTALLATION LABOR COSTS

FORM NO. 19

The data on this schedule is identical to that in Form No. 17, except that dollar expenditures are entered rather than paid man-hours. Data is normally rounded to the nearest hundred or thousand depending on the size of the job. Appropriate indication of digits dropped should be made under the title block.

(1) SUBJECT MONTHLY INSTALLATION LABOR COSTS

(2) JOB NAME CORAL SANDS HOTEL

(3) CUSTOMER S.D. CLARKE, INC.

(4) JOB NUMBER 442

(5) PREPARED BY WRF (6) APPROVED BY CAS (7) DATE 10/4/77

(8) CHANGE NUMBER 1

(9) START DATE 2/78

(9) COMPLETION DATE 4/79

(15)	(10) Y	ITD	(12) 1978												1979			
	(11) M	—	J	F	M	A	M	J	J	A	S	O	N	D	J	F	M	A
10–100 CONDUIT & RACEWAY	P	(13)	—	1.0	5.0	4.0	4.0	10.0	8.0	8.0	5.0	2.0	2.0	1.0	—	—	—	—
	A	(14)																
-200 WIRE & CABLE	P	—	—	—	—	2.0	4.0	5.0	4.0	4.0	5.0	4.0	3.0	2.5	2.0	2.0	1.5	—
	A																	
-300 DISTRIBUTION EQUIPMENT	P	—	—	—	1.0	4.0	4.0	5.0	4.0	4.0	2.5	2.0	1.0	1.5	1.0	—	—	—
	A																	
-400 FINISH WORK	P	—	—	—	—	—	—	—	—	—	2.0	4.0	8.0	10.0	8.0	8.0	10.0	3.0
	A																	
-500 SPECIAL SERVICE SYST.	P	—	—	—	—	—	2.0	10.0	12.0	12.0	20.0	16.0	16.0	20.0	12.0	12.0	15.0	3.0
	A																	
-600 MOTOR & CONTROLS	P	—	—	—	—	—	—	1.0	2.0	8.0	10.0	8.0	8.0	5.0	4.0	1.5	0.5	—
	A																	
-700 GROUNDING & BONDING	P	—	—	—	—	—	2.0	5.0	4.0	4.0	7.0	8.0	8.0	5.0	3.0	2.0	1.0	—
	A																	
-800 OUTSIDE DIST. SYST.	P	—	—	—	—	—	—	—	—	—	—	4.0	16.0	20.0	16.0	5.0	2.0	1.0
	A																	
-900 OTHER	P	—	—	1.0	1.0	1.0	1.0	1.0	0.5	0.5	0.5	0.5	0.5	0.5	—	0.5	—	0.5
	A																	
	P																	
	A																	
TOTAL	P			2.0	7.0	11.0	17.0	37.0	34.5	40.5	52	48.5	62.5	65.5	46	31	34.5	7.5
	A																	

Form No. 19: Monthly Installation Labor Costs

MONTHLY OTHER DIRECT JOB COSTS

FORM NO. 20

The explanations below are keyed in the same manner as the Installation Schedule, Form No. 11.

(1) Enter the title "Monthly Other Direct Job Costs."

(2) through
(11) Same as Installation Schedule, Form No. 11.

(12) Inception to date column, not used on new jobs. On revised plan, enter ITD costs opposite P.

(13) Enter planned monthly expenditures under the appropriate month. Figures are normally rounded to the nearest hundred or thousand depending upon the size of the job. Appropriate indication should be made under the title block.

(14) Used for recording actual costs.

(15) Enter the standard intermediate cost breakdown for Other Direct Job Costs as indicated on the sample schedule or as set forth in chapter five.

(16) Enter "30-100 Subcontract Committed" in blank heading column block and enter planned subcontract commitments under appropriate months.

(17) Enter "30-100 Subcontract Costs" in blank heading column block and enter planned subcontract payment schedule based on anticipated subcontract completion schedule.

(1) SUBJECT MONTHLY OTHER DIRECT JOB COSTS

(2) JOB NAME CORAL SANDS HOTEL

(3) CUSTOMER S.D. CLARKE, INC.

(4) JOB NUMBER 442

(5) PREPARED BY WRF (6) APPROVED BY CAS (7) DATE 10/4/77

(8) CHANGE NUMBER 1

(9) START DATE 2/78

(9) COMPLETION DATE 4/79

(15)	(10) Y	ITD	(12)	1978											1979			
	(11) M	—	J	F	M	A	M	J	J	A	S	O	N	D	J	F	M	A
40-100 SUPERVISION	P	(13)	—	1.0	2.0	2.0	2.0	2.5	2.0	2.0	2.5	2.0	2.0	2.5	2.0	2.0	2.5	1.5
	A	(14)																
-200 P/R TAXES	P	—	—	0.2	0.4	0.5	0.6	1.0	1.0	1.0	1.3	1.2	1.5	1.6	1.2	1.1	1.0	0.5
	A																	
-300 EQUIPMENT	P	—	—	0.1	0.2	0.3	0.3	0.5	0.5	0.5	0.6	0.6	0.8	0.8	0.6	0.6	0.5	0.3
	A																	
-400 SMALL TOOLS	P	—	—	0.1	0.1	0.1	0.1	0.2	0.3	0.3	0.3	0.3	0.4	0.4	0.3	0.3	0.2	0.1
	A																	
-500 MISCELLANEOUS	P	—	—	1.0	1.0	1.0	1.0	2.0	2.0	2.0	2.0	2.0	2.0	2.0	2.0	1.0	1.0	1.0
	A																	
TOTAL	P	—	—	2.4	3.7	3.9	4.0	6.2	5.8	5.3	6.7	6.1	6.7	7.3	6.1	5.0	5.2	3.4
	A																	
(16)	P																	
	A																	
30-100 SUB. CONT. COMM.	P	—	—	—	—	—	—	—	—	—	—	—	—	—	—	—	—	—
	A																	
(17)	P																	
	A																	
30- SUB. CONT. COSTS	P	—	—	—	—	—	—	—	—	—	—	—	—	—	—	—	—	—
	A																	
	P																	
	A																	

Form No. 20: Monthly Other Direct Job Costs

CUMULATIVE OTHER DIRECT JOB COSTS
FORM NO. 21

The explanations below are keyed in the same manner as for the Installation Schedule, Form No. 11.

(1) Enter the title "Cumulative Other Direct Job Costs."

(2) through
(9) Same as Form No. 11.

(10) Legend block. Enter data as indicated in sample chart:
Dashed line indicates plan;
Solid line indicates actual expenditures.
The same headings are entered in the lower left-hand corner and the planned cumulative expenditures are entered opposite plan. The data is developed from the Monthly Other Direct Job Costs Schedule (chapter three).

(11) Enter months and years, separating years by inserting vertical lines at the appropriate places.

(12) Plot plan line using appropriate coded line.

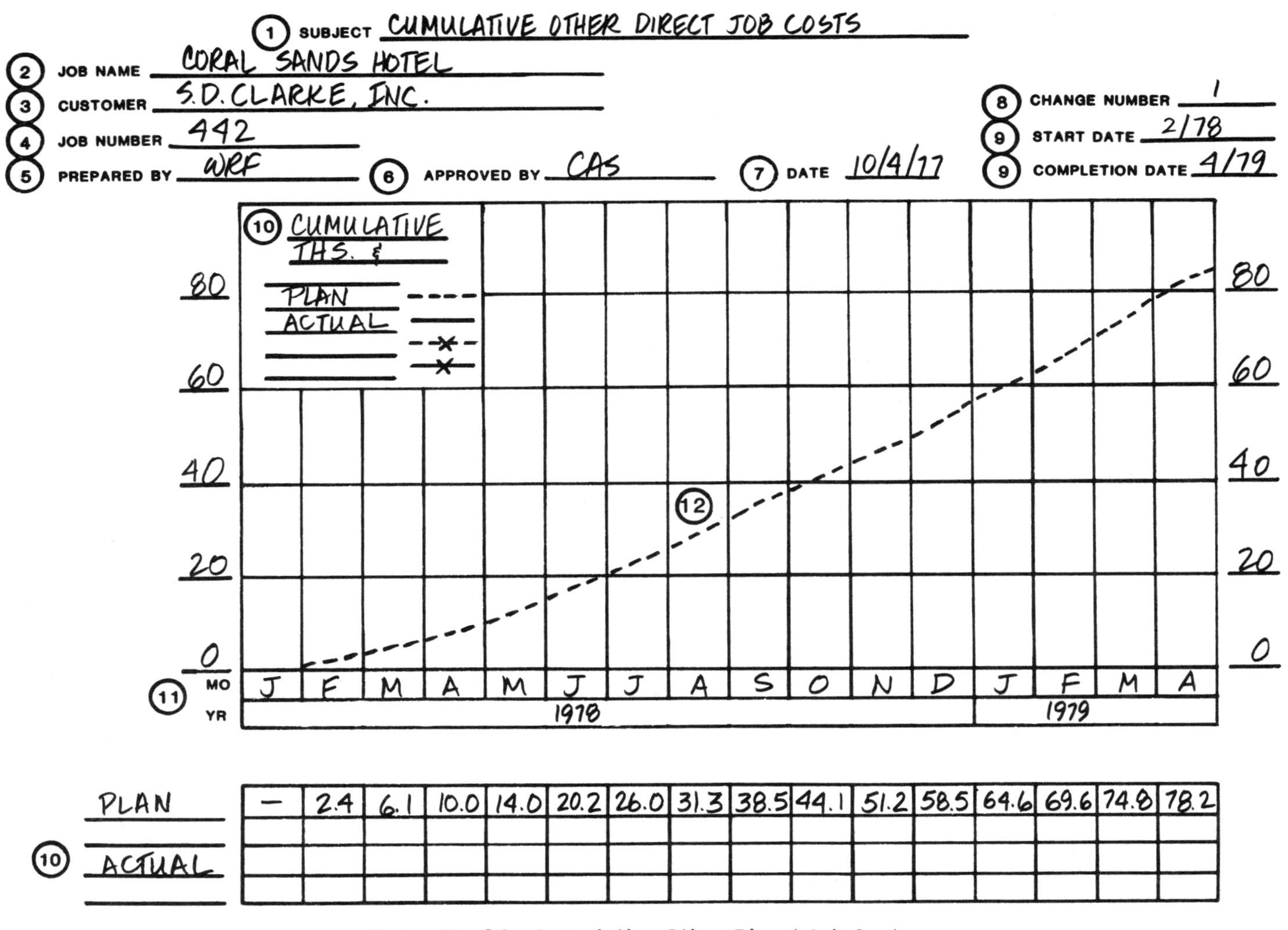

10	J	F	M	A	M	J	J	A	S	O	N	D	J	F	M	A
PLAN	—	2.4	6.1	10.0	14.0	20.2	26.0	31.3	38.5	44.1	51.2	58.5	64.6	69.6	74.8	78.2
ACTUAL																

Form No. 21: Cumulative Other Direct Job Costs

SUMMARY PLAN—CUMULATIVE

FORM NO. 10

The explanations below are keyed in the same manner as for the Installation Schedule, Form No. 11.

(1) Enter the title "Summary Plan—Cumulative."

(2) through
(11) Same as Installation Schedule, Form No. 11.

(12) Inception to date column is not used on new jobs. On revised plan, enter ITD cost opposite P.

(13) Enter planned cumulative expenditure or commitments under the appropriate month as indicated in the column to the left.

(14) Used for recording actual costs and commitments.

(15) Enter summary plan headings in the heading column spaces as outlined below and enter the required data from the planning schedule indicated:

Heading	Source
a. Material Commitments	Cumulative Material Committed & Received Chart
b. Material Receipts	
c. Subcontract	- Monthly Other Direct Job Costs—Schedule 30-100 Subcontract Costs Cumulated
d. Labor Hours	- Cumulative Man-hours Paid Chart
e. Labor Dollars	- Monthly Installation Labor Costs Cumulated
f. Other Direct Costs	- Cumulative Other Direct Job Costs Chart
g. Total Direct Costs	- Total of b, c, e & f above
h. Physical Percent Complete	Completion Status Chart Cumulative Percent
i. Billing Percent	

(1) SUBJECT SUMMARY PLAN - CUMULATIVE

(2) JOB NAME CORAL SANDS HOTEL

(3) CUSTOMER S.D. CLARKE, INC.

(4) JOB NUMBER 442

(5) PREPARED BY WRF (6) APPROVED BY CAS (7) DATE 10/4/77

(8) CHANGE NUMBER 1

(9) START DATE 2/78

(9) COMPLETION DATE 4/79

(15)	(10) Y	ITD	(12) 1978												1979			
	(11) M	—	J	F	M	A	M	J	J	A	S	O	N	D	J	F	M	A
MATERIAL COMM.	P	(13)	460	520	530	540	550	560	565	570	575	580	585	590	593	596	597	601
	A	(14)																
MATERIAL RECEIVED	P	—	10	70	90	140	195	230	265	340	430	455	475	515	563	596	601	603
	A																	
SUB. CONT.	P	—	—	—	—	—	—	—	—	—	—	—	—	—	—	—	—	—
	A																	
LABOR HOURS	P	—	—	0.4	1.8	4.0	7.4	14.8	32.1	33.4	40.6	50.3	67.6	75.1	84.4	96.7	99.4	102.0
	A																	
LABOR $	P	—	—	2.0	9.0	20.0	37.0	74.0	108.9	141.0	301.0	249.5	311.0	376.5	422.1	464.6	497.9	511.1
	A																	
OTHER DIRECT COSTS	P	—	—	2.4	6.1	10.0	14.0	20.2	26.0	31.8	38.5	41.6	51.2	58.5	64.6	69.6	74.8	78.2
	A																	
TOTAL DIRECT COSTS	P	—	10.0	74.4	105.1	170.0	246.0	324.2	399.5	520.8	659.5	749.1	837.2	950.0	1055	1130	1171	1180
	A																	
	P																	
	A																	
PHYSICAL % COMPLETE	P	—	—	—	2	4	7	14	22	29	38	47	61	73	84	89	95	100
	A																	
BILLING %	P	—	—	1	7	10	16	25	33	39	52	64	75	79	88	97	99	100
	A																	
TOTAL	P																	
	A																	

Form No. 10: (repeated): Summary Plan—Cumulative

DAILY JOB DIARY

FORM NO. 22

The Daily Job Diary Form is completed each day. Copies are distributed as follows:

a. Original—Field Office File.

b. Copies to division contracts manager or area office manager. The division contracts manager or area office manager makes distribution as deemed necessary by division chief estimator and retains a copy in the job file.

The Daily Job Diary is completed by the job superintendent as events occur during the day or as soon as possible before the end of the day to insure accuracy and completeness. A blank form is attached. The numbers below are keyed to numbers encircled on the form and set forth the requirements for the completion of each item on the form:

(1) Write in the name of the job as indicated on the approved Budget Request form.

(2) Write the name of the owner.

(3) Write in the job number as indicated on the Budget Request form.

(4) The first Daily Job Diary report completed is serially numbered 001. Each succeeding report is numbered serially 002, 003, 004, etc. This number is entered here.

(5) Indicate the date of the report.

(6) Indicate the start and completion dates for the job.

(7) Indicate the midday temperature and check status of weather.

(8) List all visitors by name and indicate the company they represent and the purpose of the visit.

(9) Indicate the number of men on the job by classification and in total.

DAILY JOB DIARY

(1) JOB NAME: ______________________ (3) JOB NO. ________

(2) CUSTOMER: ______________________ (4) SERIAL NO. ________

START DATE: ________ (6) COMPLETION DATE: ________ (5) DATE: ________

(7) WEATHER: GOOD ____ TEMPERATURE ____ RAIN ____ SNOW ____ FOG ____

(8) VISITORS AND PURPOSE ______________________

____

(9) NUMBER OF MEN ON JOB: ELECTRICIANS–J ____ F ____ G.F. ____ APPR. ____ SPL. ____ LABORERS ____

NON–CRAFT SUPPORT ____ OTHER ____ TOTAL ____

(10) IS PROJECT ON SCHEDULE? ☐ YES ☐ NO COMMENT: ________

COST IMPACT: ________ REMEDY: ________

(11) IS MATERIAL ON JOB ADEQUATE? ☐ YES ☐ NO COMMENT: ________

COST IMPACT: ________ REMEDY ________

(12) ARE TOOLS & EQUIP. ADEQUATE? ☐ YES ☐ NO COMMENT: ________

COST IMPACT: ________ REMEDY: ________

(13) IS MANPOWER ADEQUATE? ☐ YES ☐ NO COMMENT: ________

COST IMPACT: ________ REMEDY: ________

(14) WAS THERE A CHANGE IN SCOPE? ☐ YES ☐ NO COMMENT: ________

COST IMPACT: ________

(15) MEETING RESULTS & GENERAL COMMENTS: ________

____

______________________ ______________________

"NO" ANSWERS ALWAYS REQUIRE COMMENT PROJECT MANAGER / SUPERINTENT

Form No. 22: Daily Job Diary

(10) Are you meeting the planned schedule? If no, indicate areas of variance, the possible cost impact, the reasons for the variance, and the corrective measures being taken.

(11) Are you having any material problems? If so, indicate your shortages or problems, the possible cost impact, and corrective action being taken.

(12) If you have tooling or equipment problems so indicate and spell out the possible cost impact and action being taken for solution.

(13) If manpower is not adequate, indicate the possible cost impact, improvements which are necessary, and what is being done to improve the situation.

(14) Have you been given any new directions by the owner? If so, indicate the nature, probable scope, the possible cost impact, and the actions you are taking to secure and insure proper contractual coverage.

(15) Describe all meetings relating to the job including the attendees, the subject of the meeting, and the agreements and results attained by the meeting. Attach meeting minutes if available. Also add any other comments concerning the project that should be recorded to provide a complete record of activity on the job.

WEEKLY SUMMARY OF JOB PROGRESS—MAN-HOURS

FORM NO. 4

The Weekly Summary of Job Progress—Man-Hours is prepared on the first day of each week covering the man-hours expended through the previous week. The original is retained by the division controller or area office manager and copies are distributed as follows:

1st (yellow) copy to corporate director of finance;

2nd (blue) copy to area or project managers;

3rd (pink) copy to the job superintendent to be maintained in the job file.

The form is processed as follows:

Division or area accounting forwards the completed report for the prior week (pink copy) to job superintendent together with a blank form to be completed for the current week.

On Friday of each week the job superintendent completes Column 4, Estimated Percent Complete, and on the last week of each month or when a significant change occurs, he completes Column 12, Estimate at Completion—Total, and provides significant comments on the week's activity or reasons for a change in the Estimate at Completion. The form with data properly entered is forwarded to the area or division accounting.

Area or division accounting completes and issues the report.

A completed sample form is provided. The numbers are keyed to the numbers encircled on the face of the sample and indicate the data required and the method of development:

(1) Write in the sequence number of the report, starting the first report with the number one (-001).

(2) Write the name of the job as indicated on the approved Budget Request.

SUBJECT *Monthly Summary of Job Progress*

(2) JOB NAME ______ (1) REPORT NUMBER ______

(3) CUSTOMER ______ (15) PROGRESS BILLING: ______ (5) WEEK ENDING ______

(4) JOB NUMBER ______ (7) START DATE ______

PREPARED BY ______ (6) CHANGE NUMBER ______ (7) COMPLETION DATE ______

(8)		1 FIRM BUDGET HOURS	2 PRELIM. BUDGET HOURS (9)	3 TOTAL BUDGET HOURS	4 EST. % COMP (10)	5 HOURS EARNED	HOURS PAID (13)						ESTIMATE AT COMPLETION (14)	
							WEEK (11)		INCEPTION TO DATE (12)		ITD (0)/UNDER HOURS EARNED			
							6 TOTAL	7 OVER-TIME	8 TOTAL	9 OVER TIME	10 HOURS	11 %	12 TOTAL	13 (1)/U BUDGET
10-100	CONDUIT & RACEWAYS													
-200	WIRE & CABLE													
-300	DISTRIBUTION EQUIPMENT													
-400	FINISH WORK													
-500	SPECIAL SERVICE SYSTEMS													
-600	MOTORS & CONTROLS													
-700	GROUNDING & BONDING													
-800	OUTSIDE DISTRIBUTION													
-900	OTHER PRODUCTIVE COSTS													
	TOTAL													

(16) EXPLANATION OF VARIANCE & COMMENTS:

Form No. 04 (repeated): Weekly Summary of Job Progress—Tenant Work

(3) Write in the name of the owner.

(4) Write in the assigned job number per the approved Budget Request form.

(5) Indicate the date of the last day of the week for which the report is being prepared.

(6) Indicate the change number from the last approved Budget Request form.

(7) Indicate the start and completion dates for the job.

(8) The report is broken down according to the intermediate level of the standard cost account breakdown system for electrical construction work.

(9) Accumulate the budget hours from the approved Budget Request for the basic contract and for each approved firm change. Enter the total of these budget hours in Column 1 opposite the appropriate breakdown description. Accumulate the preliminary budget hours on all change order budgets marked "Preliminary" which have not been succeeded by firm budgets and enter in Column 2 opposite the appropriate breakdown heading. Total Column 1 and Column 2 to arrive at the total budget hours and enter in the appropriate block in Column 3.

(10) At the close of each week, inventory the job and estimate the percentage of work complete in each category of work for which a budget has been established. Enter the estimated percent complete in the appropriate space in Column 4. Multiply the estimated percent complete times the budget hours to compute hours earned and enter in Column 5. To compute the total percent complete, divide the total budget hours into the total earned hours as computed above. Enter the total estimated percent complete in Column 4 opposite the total.

(11) Enter weekly total hours paid including overtime hours in Column 6 entitled "Total." Enter overtime hours worked in Column 7 entitled "Overtime." This data is accumulated from the Daily Labor Distribution form.

(12) Inception to date hours paid are computed by adding the current week total and overtime hours to the inception to date totals from the prior week's report. These totals are entered in Column 8 for total hours and Column 9 for overtime hours.

(13) Subtract inception to date total hours Column 8 from hours earned Column 5 and enter in ITD (0)/Under Hours Earned, Column 10. Divide Column 10 by Column 5 and enter in Column 11 ITD (0)/Under Hours Earned Percentage.

(14) Enter your best estimate of the total hours paid that were required to complete all work budgeted under each item (hours paid to date plus hours paid to complete = hours paid at completion) and enter in Estimate at Completion Column 12 entitled "Total." In Column 13 entitled "(0)/Under Budget" enter the difference between the estimated total hours paid at Completion Column 12 and total budget hours Column 3.

(15) When a new progress billing percentage is approved by the appropriate customer representative having authority to negotiate billing status, enter the agreed upon percent amount in this block.

(16) To provide a comprehensive analysis of the week's activities, provide comments on any significant variance either in hours paid versus hours earned, or estimate to complete. Also provide comments on significant activities and accomplishments during the week.

Administration and Processing of Changes and Backcharges

INTRODUCTION

A change may be defined as any express or implied request for material or work which modifies, increases, or decreases the original scope of work contained in the base contract. An example of a change which involves an increase to the scope of work would be the addition of a new emergency generator to a hospital renovation project on which the customer had originally specified that alterations to the electrical system be adapted to the existing emergency generator. A deletion from the scope of work would occur if the customer involved in a high rise office building project decided to reduce the number of floors, thereby eliminating a proportionate amount of the secondary electrical service.

The scope of work may also be modified without changing the quality or quantity of the material which is ultimately to be installed. For example, the original contract documents may indicate that conduit is to be installed below existing piping and ductwork on a renovation project. If the customer later requests that the conduit be installed above the existing mechanical work in order to increase the space between the floor and ceiling, this constitutes a change to the original scope of work even if there is no increase in the amount or type of conduit required, since it increases the labor costs of installing the conduit by making access to the area of installation more difficult.

Although construction contracts invariably give the customer the right to order changes in required performance in return for equitable compensation, some changes are so significant and substantial that they cannot be required by the customer. These changes are often referred to as "cardinal" changes. They arise when the customer requires the electrical contractor to perform work outside the general scope of work of the base contract. Unlike an ordinary change proposal, a cardinal change constitutes a breach of contract and, therefore, discharges your duty to perform this work. An example of a cardinal change would be a requirement that the contractor build an entirely different building based on a set of plans and specifications other than those in the original contract documents. Because such extreme examples are rare, and because the consequences of a breach of contract are harsh, a determination that a cardinal change has been ordered should not be made without the advice of your consultant or counsel.

Sometimes the customer orders a change without admitting that it is a change. This action constitutes a "constructive" change and gives you all the rights you would have had if the customer had recognized the directive as a formal change order. Because the existence of a constructive change implies that a dispute has already arisen, a constructive change ordinarily must be resolved through claim negotiation, arbitration, or litigation.

Backcharges are closely related to changes, since they generally result when one party to a contract performs work that a second party should have completed and which was therefore beyond the first party's original scope of work. Typically, backcharges are issued before any formal documents, such as change orders authorizing the work, are executed. On the one hand, you may backcharge your customer, subcontractors, or other trades on the project by invoicing the party which caused you to perform the work. On the other hand, backcharges may be issued against you by your customer, subcontractors, or other trades if any of those parties performs your work because you authorized them or caused them to do so.

POLICY AND PURPOSE REGARDING CHANGES AND BACKCHARGES

Certain contract administrative controls are required for the preparation, negotiation, and execution of changes and backcharges from the time of their initiation through their final disposition. The purposes of these controls are to provide a coordinated effort to assure that changes and backcharges are properly executed and to effect the timely recovery of costs, at a minimum expense, for the work performed in connection with them.

It must be recognized that changes are often advantageous to your company. This situation exists because change orders are negotiated in a relatively noncompetitive atmosphere. The mobilization and coordination costs which would result from the customer going to one of your competitors to perform changed work give your company a tremendous advantage in pricing its change orders.

Conversely, changes which delete items from the scope of work may be viewed as opportunities to eliminate certain risks inherent in your company's original estimate.

Changes, however, do not necessarily always benefit your company. For example, major alterations to the scope of work may cause your corporation to incur significant disruption costs which are difficult to quantify, prove, and recover. In another instance, changes implemented near the time of project completion under a contract which only allows a small amount of mark-up for overhead and profit may be extremely costly if they force you to maintain a job site office for a substantially increased time beyond that stated in the original plan.

PREPARATION, NEGOTIATION, AND EXECUTION OF CHANGE ORDERS

The project manager should be assisted by the contract administration department or contract manager in carrying out the following procedures regarding changes. These procedures begin with the identification of a change to the contract based upon a regular review of project correspondence, project reports, and site investigations. When a change is identified, its existence should be documented immediately by an appropriate letter to the customer. This correspondence need not be a formal change order proposal; a breakdown of the costs which are claimed for the change may be forwarded later, since the primary purpose of this correspondence is to notify the customer of the existence of the change. It is also possible that the change will be initiated by the customer, in which case the customer's request for a proposal will adequately document the existence of the change.

Once the change is documented, you must develop an estimate of the time and cost of completion of the involved work just as you did for the base contract work when the project was originally bid or negotiated. Change orders are generally priced in three basic ways: by a stipulated lump sum price, by unit prices, or on a time and material or cost-plus basis. The method for pricing change orders is generally stated in the Changes article of your contract.

Regardless of the type of pricing used in the proposed change order, it is essential that all prices and quantities for material, all wages, and all work units which are used in calculating the price of the change order be completely accurate and supportable. If they are not, you will lose your credibility with the customer and will find it much more difficult to negotiate future change orders.

Your overriding goal in preparing and negotiating change orders should be to obtain the most favorable terms possible given the available supporting documentation. In the case of materials, you need not give bulk discounts such as those which are used in bidding a base contract, since change orders are negotiated and executed on a piecemeal basis. This is particularly true when materials not specified in the base contract are to be supplied on a unit price basis. In regard to labor cost calculation, it is essential that fringe benefits, payroll taxes, insurance, and overtime premiums, as well as the base labor rate, be included in any estimated costs for performing the change. The final cost item which must be included in all proposed change orders is the mark-up for overhead and profit.

In addition to all the direct costs for the changed work, a proposed change order should request appropriate time extensions and reimbursement for the costs which will result from the impact of the change on base contract work. Since time extensions and impact costs are difficult to compute, two possible approaches to negotiating these items are available to your corporation. First, you may rely on a relatively strong bargaining position, and on the customer's difficulty in challenging computations for time extensions and impact costs by making an accurate assessment for these items in the proposed change order. Second, you may want to require that a clause reserving your rights to claim impact costs and time extentions be included as part of the final change order. For example, you should consider using the following reservation of rights in change order proposals and in change orders themselves:

> Our proposal is based solely upon the usual cost elements such as labor, material, and normal mark-ups and does not include any amount for additional changes in the sequence of work, delays, disruptions, rescheduling, extended overhead, overtime, acceleration and/or impact costs, and the right is expressly reserved to make claim for any or all of these related items of cost prior to the settlement of this contract.

As a practical matter, many customers will be reluctant to sign a change order with such language included on the face of the document—and some will adamantly refuse to do so. Under these circumstances, you should consider proposing, as an alternative, that the customer insert in the change order this language:

> This change only covers the direct costs incurred by the contractor in performing the work which is the subject hereof.

While such language does not expressly reserve your right to seek a time extension or impact costs relating to the change to your work, nevertheless, you can argue that this reservation of rights was the intended implication since the language specifically states the change order is only covering your "direct costs."

Certain other qualifications to the method of calculating change order prices should always be considered. For instance, to avoid unanticipated labor and material cost escalations, you should make your change proposals contingent upon acceptance by the customer within a certain expressly stated number of days, or upon acceptance while prices or wages remain at a specified level. Similarly, you should clearly indicate when change proposals are dependent upon the performance of work in a certain sequence. For example, if a change proposal to install conduit in a pipe chase is based upon the assumption that the mechanical contractor will have finished his work in the same chase before installation of the conduit begins, that assumption should be expressly stated in the proposed change order.

Credits for the deletion of work also require careful attention and treatment. For example, the customer should not receive a credit for the total cost of the deleted work if certain aspects of the work, such as the preparation of shop drawings, or the fabrication of materials, have already been completed.

In the preparation and negotiation of proposed change orders, you must take into account certain general considerations as well as the specific procedures set forth in this section. The general attitude of the customer in negotiating changes, the urgency of the customer's need for the change, the relative lack of competition, and the customer's awareness of all of these factors, must be taken into account in preparing and negotiating any change order.

After a proposed change order has been transmitted to the customer, negotiated, and finalized, it is essential that you obtain a final executed copy which authorizes in writing the changed work before any work is begun. At that time, your company should also revise its plan as necessary to reflect the effects of the change order.

Any revisions to the plan should be forwarded to the contract administration department or contracts manager along with the schedule for completing the change order work itself. The project manager and the contract administration department or contracts manager should then follow-up to see that the change order work is completed and invoiced, along with regular progress payment requisitions, in a timely and complete manner.

As an aid to your analysis of change orders, the following checklist details items which should be considered in preparing change orders.

CHANGE ORDER CHECKLIST

The following cost items should be included in the preparation of any change order:

1. Supervision*
2. Engineering
3. Labor
4. Fringe benefits for labor
5. Actual cost of material
6. Subcontracts (electrical, painting, insulation, etc.)
7. Taxes and insurance on labor
8. Taxes on material
9. Any other taxes (Use, Gross, City, etc.)
10. Rented equipment
11. Gas, oil, and any other expendable materials (welding rod, oxygen, acetylene, etc.)
12. Small tools
13. Overhead
14. Profit
15. Bond
16. Any other true cost involved (phone calls, freight, travel, etc.)
17. Delay

Special attention should be given to the following items before accepting a change order.

1. Will change order cause an overall delay of job?

---

*This is *field supervision* and is not covered by overhead, Item 13. Overhead pertains to home office only.

2. Will change order interfere with planned work and cause hidden cost?

3. What are the additional costs and expenses caused by the need for the change, such as time spent by the superintendent and foreman trying to solve the problem that leads to the change order?

4. When submitting the price for the change order, always ask for a time extension. *Never* sign a change order that has no provision for an extension of time (except as approved by the home office.)

5. If coupled with additional work, will change order cause a delay or cause a disruption of planned work schedule or equipment spread?

6. If there is a chance that the change order might cause a delay as stated in 5, then you should insist that you will accept the extra work order or sign the change order only upon the condition that the matter of time extension and extra cost be expressly excluded from the order or reserved to you for future action or claim. The above should not be a verbal agreement, but should be in writing and brought to the attention of the home office.

7. If a change order is issued under the conditions stated in 6, then the superintendent will have to keep detailed records of cost.

8. All change orders, where practicable, should be reviewed by the home office.

9. All change orders should be numbered in sequence and each different change order should carry a different number.

10. All purchase orders, subcontracts and invoices should carry this number when issued or being approved.

11. Change order cost must not become a part of the original contract cost and must be kept separate.

In the case of backcharges which are implemented against other trades with whom there is no contract, special attention must be given by the contract administration department or contracts manager to assure prompt payment, since there is no contractual basis by which you, as the contractor, can withhold funds which are due from the party being backcharged. In such a case, the contract administration department or contracts manager must assist the project manager in determining the financial solvency of the party being backcharged, invoice that party promptly, and follow up assiduously to see that the bill is paid. If the bill is not paid promptly, the contract administration department or contracts manager must immediately advise the project manager so that no additional work will be performed by your company for the party being backcharged.

RECORD-KEEPING REGARDING CHANGES AND BACKCHARGES

As part of the change order process, a file should be developed and maintained which contains all internal and external correspondence regarding the change order, the computations involved in developing the estimate for the change order, any notes of negotiating sessions, and all formal change order documents. In addition, a separate log should be maintained which will reflect the status of all changes and backcharges. This log should assign a number to each proposed change and backcharge and should reflect the following information:

A concise description of the scope of the change;

The customer's identification number;

The date that the change is documented or the request for a change proposal is received;

The date that the change proposal is due;

The date that the change proposal is submitted;

The date of any customer action to reject or accept the proposal.

Acquisition and Control of Tools and Equipment

ACQUISITION

Before the job starts, the project manager analyzes the job to determine what tools and equipment will be required. Based on his analysis, he completes the Equipment and Tool Requisition form (Form No. 23) in order to initiate procedures which will assure the availability of the necessary tools and equipment for the job as required by the contract performance schedule.

The division manager or equivalent manager must approve the list of tools and equipment submitted by the project manager. Upon doing so, he forwards this list to the equipment manager who reviews the approved list and determines the availability of the requested tools and equipment within the company. Based on that determination, your equipment manager recommends the transfer of tools and equipment from surplus stock, the transfer of tools and equipment from other jobs, or the acquisition of additional tools and equipment. He prepares a Property Transfer form (Form No. 24) for the list of items coming from surplus inventory. When the items are available on another job, he advises the project manager at that job to transfer the available items to the job requesting those items. This is also done through a Property Transfer form. The equipment manager prepares a purchase requisition for those items which are not available from internal sources and forwards it to the division manager for approval.

The division manager then reviews the purchase requisition in order to convince himself of the necessity for the new purchase and returns the requisition to the equipment manager upon approving it.

Upon receipt of the approved purchase requisition, the equipment manager enters the approval on a monthly scroll and forwards the purchase requisition to the purchasing agent for procurement.

The purchasing agent then places the purchase order and forwards a copy of the request and a numbered property identification tag to the project manager.

Upon receipt of the tools and equipment from other jobs, surplus inventory, and new purchases, the project manager inspects all

EQUIPMENT AND TOOL REQUISITION

PROJECT: ______________________________ DATE: ________

PROJECT LOCATION: ______________________________ JOB NO.: ________

ITEM	QTY	DESCRIPTION	DATE REQ	SUPPLIED BY WAREHOUSE		P.O. PLACED BY BUYER	
				DATE	DOC. NO.	DATE	P.O. NO.

PREPARED BY ____________ WAREHOUSE ____________

PROJECT MGR. ____________ ____________ BRANCH/DIVISION MANAGER BUYER ____________

Form No. 23: Equipment and Tool Requisition

PROPERTY TRANSFER

NAME OR DESTINATION	BLDG	RM OR MS	SOURCE CODE	PROPERTY ☐ CORP. ☐ GOVT.
FROM	BLDG	RM OR MS	SOURCE CODE	DATE

REASON FOR MOVEMENT: ☐ PERMANENT MOVE ☐ LOAN ☐ CALIBRATION–REPAIR ☐ TEMPORARY MOVE ☐ SURPLUS ☐ STORAGE	NO. PCS. OR PKGS.	CONTRACT NO.

ITEM	QUAN	PROPERTY IDENTIFICATION NO.	DESCRIPTION

AUTHORIZED BY	SOURCE CODE	DATE	TRANSPORTATION	DATE

PROPERTY ADMINISTRATION	DATE	RECEIVED BY	SOURCE CODE	DATE

PROPERTY ADMINISTRATION COPY

Form No. 24: Property Transfer

items to ensure that they are in accordance with the needs of the corporation and that they are as described in the transfer form on the purchase order copy provided by the purchasing agent. The project manager notifies the equipment manager of any discrepancies, and of damaged or inoperative items received by noting such problems on the receiving document. The project manager acknowledges receipt by signing the Property Transfer form and forwarding it to the equipment manager. He then affixes the property identification tags provided by the purchasing agent, enters the tag number on the receiving notice, completes the form, and forwards it to the equipment manager.

TOOL CONTROL

Before work commences, the project manager designates a tool control man to be responsible for the issuance, maintenance, and control of all tools as well as the tool crib area. All tool transfers are made to other jobs or returned to the warehouse by completing a Property Transfer form. Throughout the performance of the work, the project manager monitors the activity of the tool control man to ensure that he is properly carrying out his duties and that he is preventing an excessive number of tools from being lost, stolen or damaged. When the job is completed, the project manager has the tool control man make a physical inventory of the tools on hand. Based on this inventory, the project manager prepares a Property Transfer form and returns the remaining tools and equipment to the company warehouse.

EQUIPMENT CONTROL

The project manager is responsible for providing proper preventive maintenance for all pieces of equipment assigned to the job. Any major repairs to equipment which are undertaken are coordinated with the equipment manager before they are accomplished. The project manager returns unneeded equipment to the company warehouse using the Property Transfer form. As in the case of the final tool inventory, upon completion of the job the project manager inventories all equipment on hand, completes the Property Transfer form and returns the equipment to the company warehouse.

Maintenance of On-Site Marked-Up Contract Drawings

As part of his contract administration responsibilities, the project manager must at all times maintain on each job site the latest set of contract construction drawings, marked-up on a current basis to provide a visual record of work in place.

In order to initiate the development of this record, the project manager secures one set of the latest issue of all contract construction drawings and supplies them to the field office prior to start up of construction on each project. These drawings are maintained in the field office at all times. As changes are implemented, revised drawings are provided in a similar manner. By noon on Monday of each week, each foreman marks up the drawings with colored pencils indicating all materials and/or equipment installed under his direction as of the end of the previous week. A legend is maintained on the drawings setting forth the colors and symbols being used for marking up the work accomplished. Once a week, the project manager reviews the marked-up drawings to ensure that they have been brought up to date and that proper markings are being used. During this audit, a spot check of the project is made to ensure that items marked-up as installed have in fact been installed and are complete.

Safety

THE IMPORTANCE OF A SAFETY PROGRAM

As an electrical contractor, you should require safe construction practices to prevent personal injuries and damage to property. Safety must be a prime consideration in all phases of construction including planning, purchasing, fabrication, construction, operation and maintenance. All practical steps must be taken to maintain safe, healthful places to work. Adequate protection equipment must be provided to minimize all foreseeable accident and health hazards. Safety is not only humane, it also has a direct effect on job profit, since accident frequency and severity determine the amount of Workmen's Compensation Insurance Premium chargeable to each job. Safety, therefore, is of no less importance than any other basic responsibility each project manager has in his goal of obtaining maximum profit possible from each job.

THE ORGANIZATION OF A SAFETY PROGRAM

The corporate safety program should be placed under the direction of the labor relations consultant or an equivalent individual. He coordinates the safety activities of field management to provide close cooperation with your Workmen's Compensation Insurance Carrier, owner safety personnel, and other safety groups. He will render advice and assistance in all matters pertaining to safety, maintain accident records, conduct investigations of and report on major industrial injuries, and supply the field with safety publications and forms.

A safety program will be established at each job under the direct supervision of the project manager. On the largest jobs, a safety engineer may be assigned responsibility for administration of the safety program under the direction of the project manager. Each job safety program will provide that:

1. Responsible supervisors shall make frequent inspections of all excavations, forms, scaffolds, stairs, ladders, structures, machinery, and equipment and shall take immediate corrective action to eliminate any hazard found.

2. Monthly meetings of all foremen will be held under the direction of the division manager for a discussion of safety problems and to review accident experience.

3. Foremen shall conduct short "tailgate" safety meetings with their crews each Monday morning on the job, to emphasize safety. Each meeting shall have a definite subject matter and a meeting report will be forwarded to Labor Relations Counsel on forms provided.

4. Each new employee shall be given instructions regarding the hazards and safety precautions applicable to the type of work he will be expected to perform and will be directed

to read the Safe Practices and Operations Code (see page 121).

5. Each job shall be supplied with an adequate First Aid Kit.
6. Contact with approved local doctors and ambulance services will be established prior to the beginning of each job.
7. A weatherproof production of the Safe Practices and Operations Code will be posted at each job location. This is mandatory under State law.

SAFE PRACTICES AND OPERATIONS CODE

The following basic rules apply to all field projects:

1. Anyone known to be under the influence of intoxicating liquor shall not be allowed on the job while in that condition.
2. Horseplay and scuffling are prohibited.
3. Work shall be well-planned and supervised to forestall injuries in the handling of heavy materials and in working with equipment.
4. No one shall knowingly be permitted or required to work while his ability or alertness is so impaired by fatigue, illness or other causes that it might unnecessarily expose him or others to injury.
5. Employees shall not enter manholes, underground vaults, chambers, tanks, silos, or other similar places that receive little ventilation, until tests have been made to determine that the air contains no flammable or toxic gases or vapors.
6. Employees should be alert to see that all guards and other protective devices are in proper places and adjusted, and shall report deficiencies promptly to the foreman.
7. Crowding or pushing when boarding or leaving any vehicle or other conveyance is prohibited.
8. Workers shall not handle or tamper with any electrical equipment, machinery, or air or water lines in a manner not within the scope of their duties, unless they have received instructions from their foreman.
9. All injuries shall be reported promptly to the foreman so that arrangements can be made for medical or first aid treatment.
10. When lifting heavy objects, use the large muscles of the leg instead of the smaller muscles of the back.
11. Shoes with thin or badly worn soles shall not be worn.

12. Do not throw materials, tools, or other objects from buildings or structures until proper precautions are taken to protect others from the falling object hazard.

13. Wash thoroughly after handling injurious or poisonous substances, and follow all special instructions from authorized sources regarding this matter. Hands should be thoroughly cleaned just prior to eating if they have been in contact with paint or similar substances.

14. Employees should avoid the use of extension ladders when carrying loads.

15. Arrange work so that you are able to face the ladder and use both hands while climbing.

16. Neither gasoline nor carbon tetrachloride shall be used for cleaning purposes.

17. No burning, welding, or other sources of ignition shall be applied to any enclosed tank or vessel, even if there are some openings, until it has first been determined that no possibility of explosion exists, and authority for the work is obtained from the foremen.

18. Any damage to scaffolds, falsework, or other supporting structures must be reported promptly to the foreman.

19. The following basic rules apply to the use of tools and equipment:
    a. Keep faces of hammers in good condition to avoid flying nails and bruised fingers.
    b. Hold cold chisels in such a way that the knuckles will be protected if the hammer misses the head. Chisels struck by others should be held by tongs or similar holding devices.
    c. Do not use pipe or wrenches as substitutes for other wrenches.
    d. Wrenches should not be altered by the addition of handle extensions or "cheaters."
    e. Files shall be equipped with handles. Never use a file as a punch or pry.

f. Do not use a screwdriver as a chisel.
g. Keep handsaws sharp.
h. Do not push wheelbarrows with handles in an upright position.
i. Do not lift or lower portable electric tools by means of the power cord. Use a rope.
j. Do not leave the cords of portable electric tools where cars or trucks will run over them.
k. In locations where the handling of a portable power tool is a problem, try handling it from some stable object, by means of a rope or similar support of adequate strength.

20. The following basic rules apply to machinery and vehicles:
a. Do not attempt to operate machinery or equipment without special permission, unless it is one of your regular duties.
b. Loose or frayed clothing, dangling ties, finger rings, etc., shall not be worn around moving machinery or other sources of entanglement.
c. Machinery shall not be repaired or adjusted while in operation, nor shall oiling of moving parts be attempted, except on equipment that is designed or fitted with safeguards to protect the person performing the work.
d. Do not work under vehicles supported by jacks or chain hoists without protective blocking that will prevent injury if jacks or hoists should fail.
e. Air hoses should not be disconnected at compressors until the hose line has been bled.

Questions:

1. Under what circumstances should the project manager revise his plan?

2. What changes need not be performed?

3. Why can changes be advantageous?

4. What are the three usual options for change order pricing?

CHAPTER FIVE

Scheduling and the Use of Scheduling Techniques

Introduction

It is now commonplace in the construction industry for contracts between owners and general contractors to require that the contractor supply, follow, and update a construction schedule. This schedule is developed according to a bar chart or a network analysis technique such as the critical path method (CPM) or precedence diagramming method (PDM). A schedule can aid all parties to a contract. It can provide the customer with an accurate assessment of the project's progress and it can allow you, as the contractor, to identify delays and to support your request for time extensions, in accordance with the area (or areas) affected. Use of these scheduling techniques allows each party to have a visual representation of the construction project.

You should not be apprehensive in using a network analysis technique, such as CPM or PDM, when requested by the customer.

Rather, you should become familiar with and understand such a scheduling technique and use it to your advantage. Schedule techniques can be particularly effective in achieving cost control; making adjustments to planned manpower, equipment, and material requirements; and scheduling of work activities. If necessary, the schedule can also be of benefit in claims presentation and defense.

Network analysis techniques, rather than bar charts, are valuable in that they (1) determine the earliest and latest allowable start and finish dates for specific activities contained in the project schedule; (2) specify the amount of float time allocated for an activity or sequence of activities; and (3) provide a method by which the schedule can be updated to allow for delays and time extensions, which exceed the respective allowable float time for each activity or sequence of activities.

This manual will limit its discussion of the bar chart and network analysis techniques to their applications and impacts on the construction project.

The Use of the Bar Chart in Scheduling

Prior to the introduction of network analysis techniques, the bar chart was essentially the only type of schedule used on a construction project. With the advancement of the network analysis techniques, the bar chart has become somewhat limited in use. The bar chart is often used on smaller projects, however, which do not require the detail available through a network analysis. The bar graph may also be used to supplement the network schedule as the bar chart is a simplified schedule which is easily understood.

Graphically, the bar chart is a series of lines, usually running horizontally, which represent items or areas of work on the project. The vertical axis represents the areas of work and the horizontal axis signifies the time-scale for the project. For a graphic illustration, see the end of this chapter.

Although the bar chart has been used for many years, with the advancement of network scheduling analyses, the disadvantages of the bar chart become apparent. One of its most advantageous aspects, simplicity, is also a disadvantage in that it does not explicitly indicate activity interrelationships, and may subsequently misinform users as to the actual working duration of an activity. For instance, an activity may actually entail an intermittant or noncontinuous performance requirement dependent upon the phasing in of other activities of work within a given area. But due to the lack of expressly designated interrelationships, the bar chart could only logically reflect a summary line from initial start to final completion.

Another problem associated with the bar graph is its inability to determine a critical path of activities on the project. It is essential in a construction project to identify which activities are critical to the completion of the project. The bar chart, however, does not provide a method for establishing which items or activities are critical.

Although the bar chart has many disadvantages, it is still effective in the scheduling of smaller projects. It can be used to supplement the network scheduling technique, because it provides a simple, easy to understand view of the start and completion dates for all the activities.

Network Techniques—Critical Path Method (CPM)

The critical path method (CPM) is a scheduling technique based on a network diagram system. Each activity is represented as an arrow with a time-scale given for each activity to represent the estimated duration. CPM provides order and logic in a schedule while emphasizing the relationship that exists with respect to the other activities contained in the schedule. The CPM schedule provides the user with the following information.

NUMERICAL IDENTIFICATION

CPM identifies the activities to be performed by numerical designations. These numbers are known as I–J numbers. The I number corresponds to a node on the network diagram which occurs at the beginning of the activity. The J number corresponds to a node on the network diagram which occurs at the end of the activity. Refer-

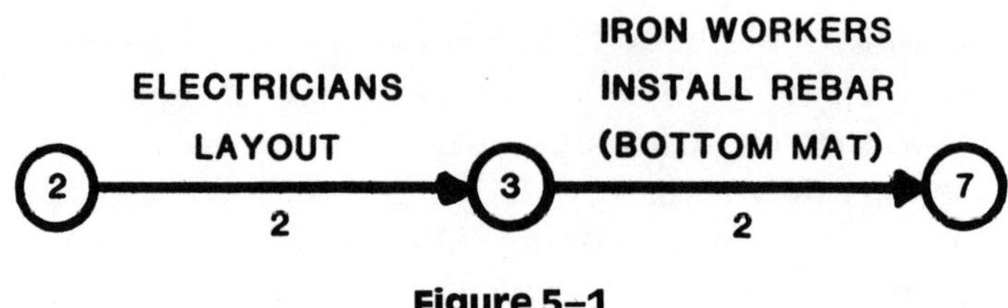

Figure 5–1

ring to Figure 5–1, taken from the Network Logic Diagram for a Typical Concrete Floor Slab supplied at the end of this chapter, the activity "electricians layout" is identified by I–J numbers as "activity 2–3." The activity "iron workers install rebar (bottom mat)" is identified as "activity 3–7" and its I number is also the J number for the preceding activity "electricians layout" activity 2–3.

ACTIVITY INTERRELATIONSHIPS

CPM shows the logic which is necessary in the performance of scheduled activities, i.e., those activities which must be complete before another activity can start.

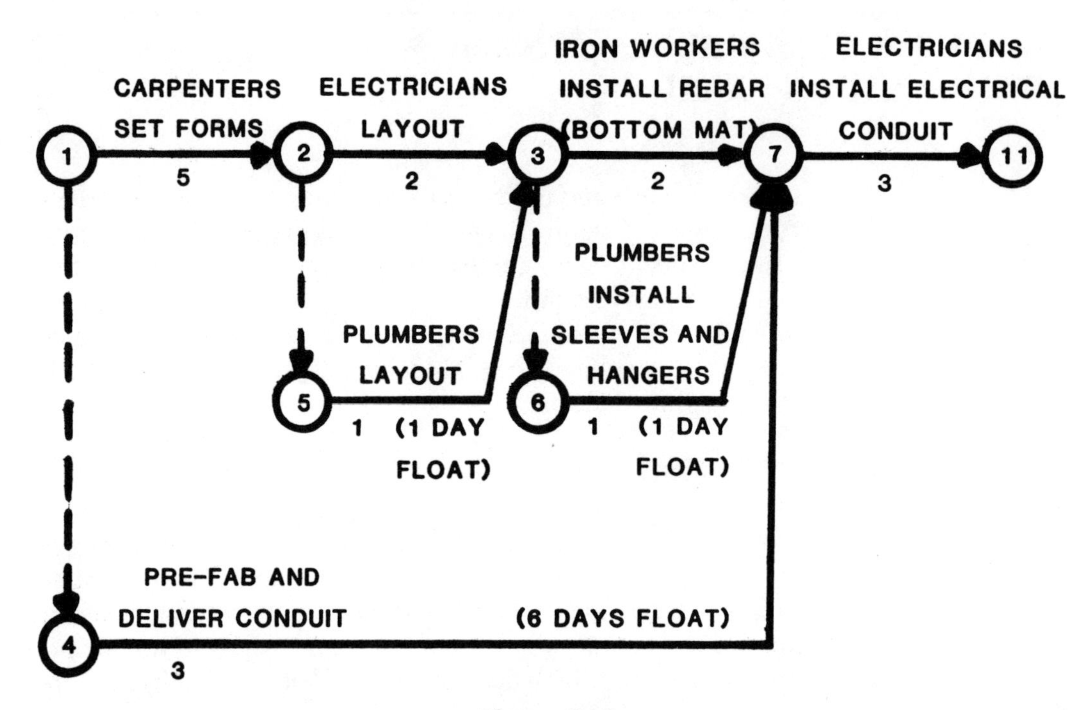

Figure 5–2

As shown in Figure 5–2, which refers again to the Network Logic Diagram for a Typical Concrete Floor Slab, the activity "electricians install electrical conduit" cannot occur until the activity "pre-fab and deliver conduit" and the chain of activities "carpenters set forms—electricians layout—iron workers install rebar (bottom mat)" is completed. However, there is no logical dependency between the activity "electricians layout" and the activity "pre-fab and deliver conduit," and therefore they can be scheduled to be performed concurrently.

DURATION AND EARLY START/EARLY FINISH SEQUENCE

CPM provides a planned duration for each activity. For example, in Figure 5–3 from the time-scaled "As-Planned" CPM Schedule at the end of this chapter, the activity "electricians install elec-

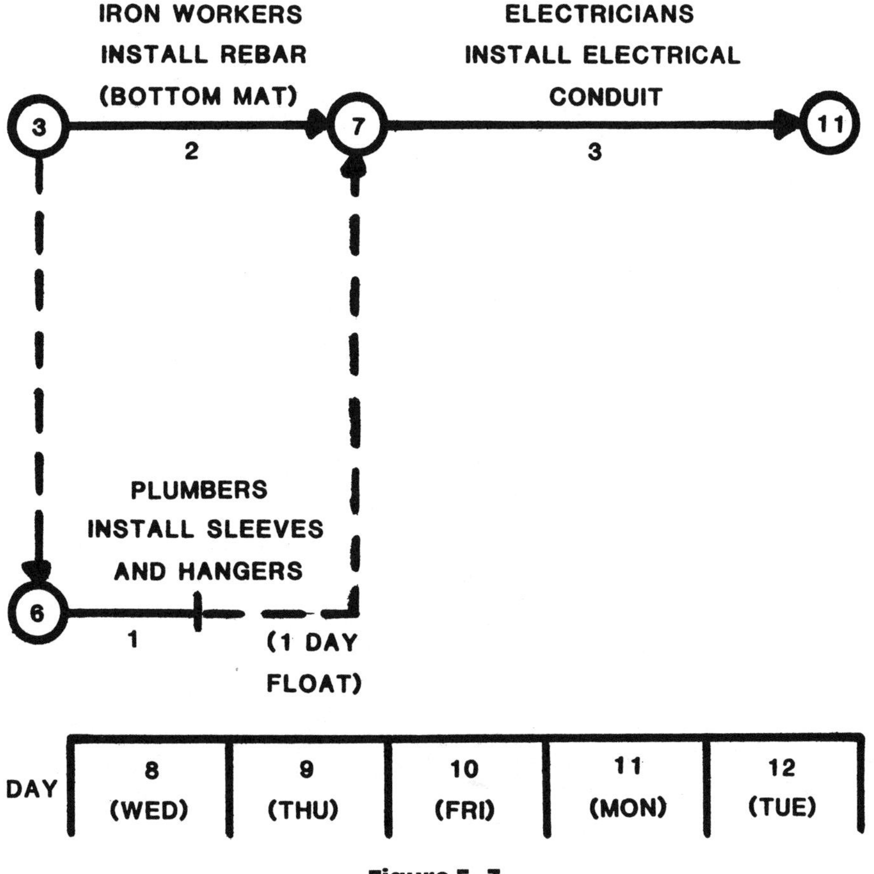

Figure 5–3

trical conduit" has a planned duration of three days from the tenth through the twelfth working day of the plan.

CPM provides early start and early finish dates for each activity. These dates represent the earliest possible start and finish for each activity given the duration of the activity and the logical constraints upon it. On the time-scaled "As-Planned" CPM Schedule, each activity is shown in the early start/early finish configuration. Thus, the activity "plumbers install sleeves and hangers" has an early start and early finish through day 8, while the activity "electricians install electrical conduit" has an early start of day 10 and an early finish through day 12.

LATE START/LATE FINISH SEQUENCE

CPM provides late start and late finish dates for each activity. These dates represent the latest possible start and finish dates which can occur without delaying overall project completion, given the duration of each activity and the logical constraints upon it.

Referring to Figure 5–4, taken from the time-scaled "As-Planned" CPM Schedule, late finish dates are established by using

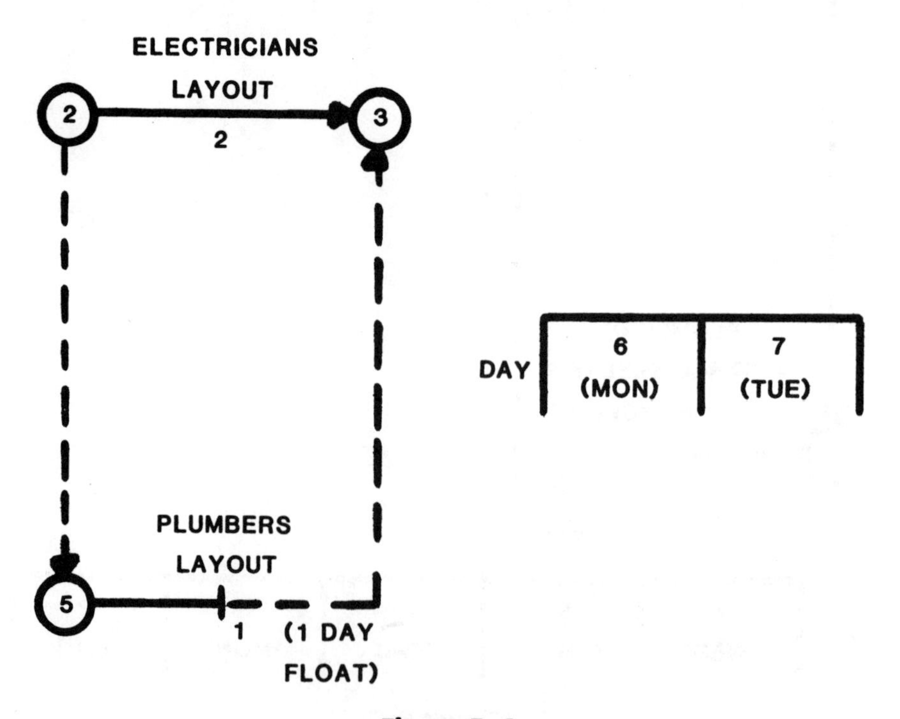

Figure 5–4

all float time available to an activity. Although late start dates are not shown, they may be calculated by subtracting the duration of each activity from the latest finish date. Performing this operation for the activity "plumbers layout" shows that the activity could start as late as day 7 in wich case its one day duration would be completed in time for the start of a subsequent activity on day 8.

FLOAT TIME

CPM shows which activities have float time and which are critical. Float time is simply that amount of time that the finish of an individual activity can be delayed without delaying the finish of the entire project. Stated differently, the differences between the early finish date of an activity and the late finish date of an activity is the amount of float. If those dates are the same, and there is no difference between them, the activity in question is on the critical path, which is simply another way of saying that it has no float. It must be recognized that float could be allocable to a series of activities rather than to an individual activity. Thus, a delay in a single activity, along a series with float, can use up the float time which was available for all other activities on that chain.

This is illustrated by Figure 5–5, which was taken from the "As-Built" CPM Schedule included at the end of this chapter. A nine day delay in the activity "architect/owner approval of shop drawings" has consumed more than the entire amount of float in the relevant chain of activities. At the same time, it has created float in the chain of activities between node 1 and node 5, shifted the critical path away from that chain of activities, and delayed the overall project.

COMPUTERIZATION OF CPM

Although CPM networks may be set forth in arrow diagrams which are time-scaled, one more commonly sees CPM networks in computer printout forms. Computerization of CPM networks provides obvious cost and time savings when it comes to updating project schedules. Since many contracts require such updates on a monthly basis, the computerization of the schedule becomes a necessity. A computer printout of a CPM differs from the arrow diagram of the same CPM in format only. It still provides the user with activities identified by activity descriptions and I–J numbers, logical relationships, activity durations, early and late start and finish dates, and float times. In addition, computer printouts may often be sorted

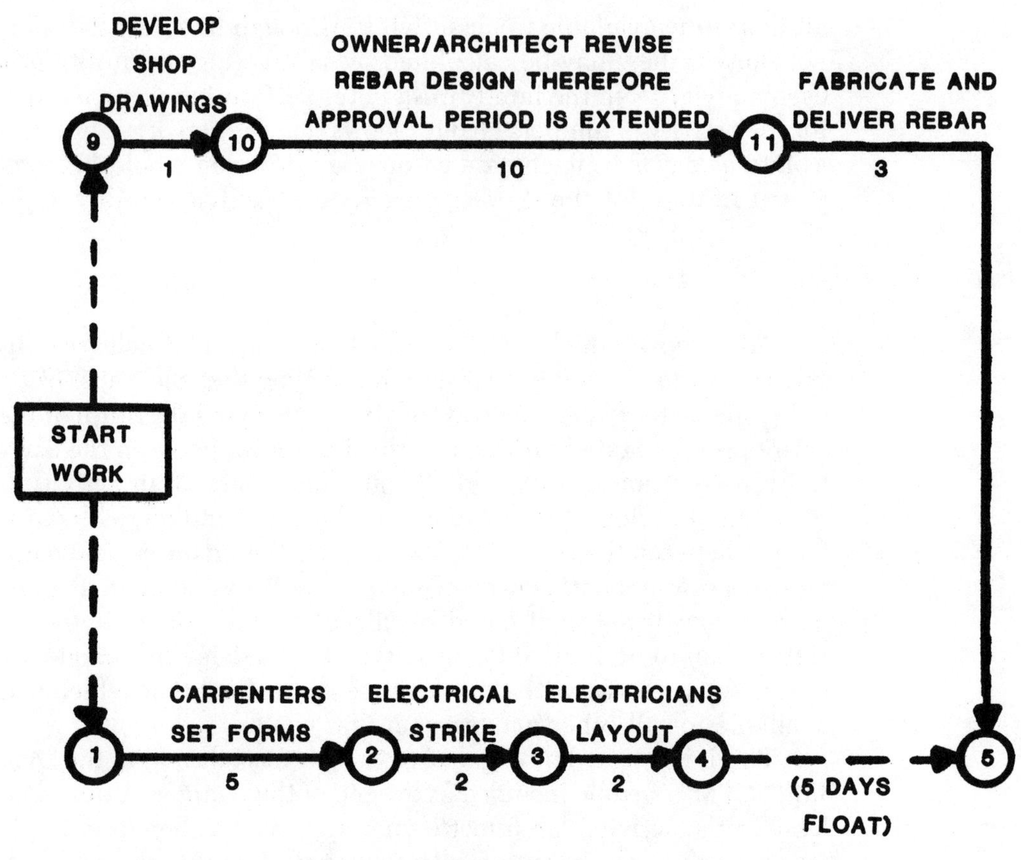

Figure 5–5

in varying fashions. For example, the computer may sort activities by the amount of float time, showing critical and near-critical activities first. The main disadvantage that the computer printout of a CPM has in comparison to a time-scaled arrow diagram is that the user cannot "see" what is happening or what has happened on the project simply by looking at the schedule.

Network Techniques—Precedence Diagram Method (PDM)

The precedence diagram method (PDM) is a scheduling technique also based upon a network diagram system. It is an activity-on-node oriented system with a time established for each activity to represent the estimated duration. PDM provides order and logic in a schedule and is capable of indicating the relationships that exist between all activities contained in the schedule. The PDM schedule provides the user with the following information.

NUMERICAL IDENTIFICATION

PDM identifies the activities to be performed by numerical designations. These numbers are known as activity numbers. As PDM is an activity-on-node oriented system, only one identification number is used for the entire activity.

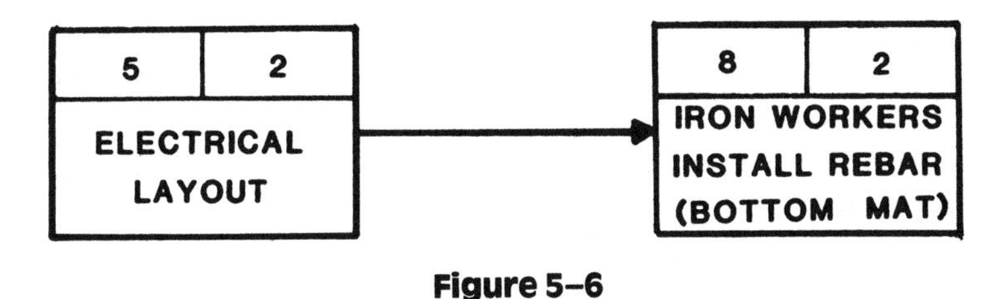

Figure 5–6

Referring to Figure 5–6, taken from the Precedence Diagram for a Typical Concrete Floor Slab, supplied at the end of this chapter, the activity "electrical layout" is identified by the number 5 in the upper left-hand corner of the box and subsequently becomes known as activity #5. The activity "iron workers install rebar (bottom mat)" is identified as activity 8 in the upper left-hand corner of the box.

ACTIVITY INTERRELATIONSHIPS

As in CPM, PDM shows the logic that is necessary to the performance of scheduled activities, i.e., those activities which must be complete before another activity can start.

As shown in Figure 5–7, which is taken from the Precedence Diagram for a Typical Concrete Floor Slab, the activity "install electrical conduit" cannot occur until the activity "pre-fab and deliver conduit" and the chain of activities commencing with "set forms" are completed. Note, however, there is no logical dependency between the activity "electrical layout" and the activity "pre-fab and deliver conduit," and these activities, therefore, can be scheduled to be performed concurrently.

DURATION AND EARLY START/EARLY FINISH SEQUENCE

The PDM provides a planned duration for each activity, which is located in the upper right-hand corner of the box. At this point, it must be noted that the precedence diagram method in its prescribed format does not lend itself to time-scaling. As illustrated in Figure 5–8, which highlights activities 8, 9, and 10 of the Precedence Diagram for a Typical Concrete Floor Slab, it may be readily observed that in order to accomplish time-scaling, the PDM system must take on the appearance of a CPM. Note that the activities are numerically identified by a single number. The activity "install electrical conduit" has a planned duration of three days from the tenth through the twelfth working day of the plan.

LATE START/LATE FINISH SEQUENCE; FLOAT TIME; AND COMPUTERIZATION OF PDM

Calculations to attain late start, late finish and float times for each activity within the precedence diagram method are handled in exactly the same manner as they would be for attaining those times within the critical path method. This rule holds true for computerization as well.

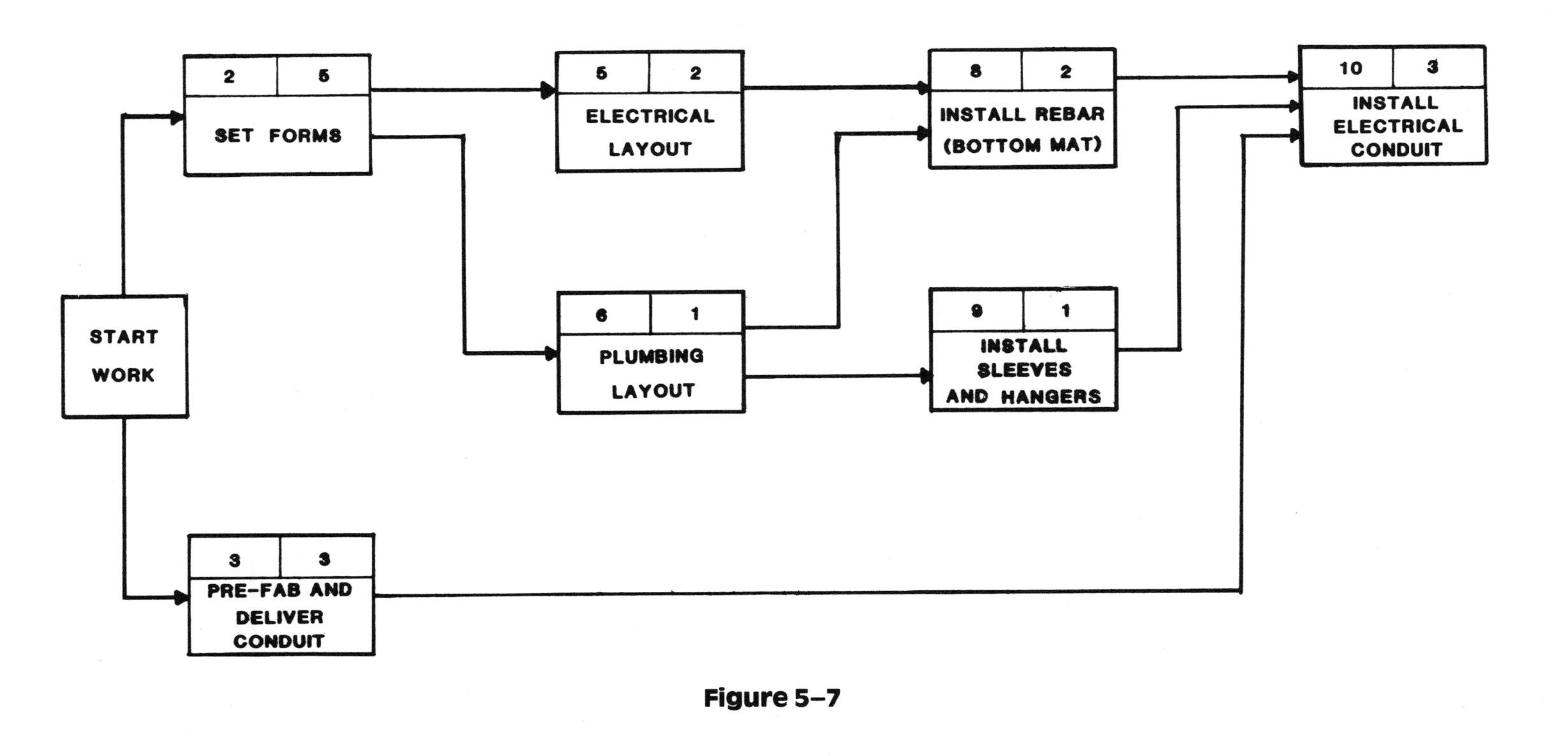
START
WORK
2
5
SET FORMS
3
3
PRE-FAB AND
DELIVER
CONDUIT
5
2
ELECTRICAL
LAYOUT
6
1
PLUMBING
LAYOUT
8
2
INSTALL REBAR
(BOTTOM MAT)
9
1
INSTALL
SLEEVES
AND HANGERS
10
3
INSTALL
ELECTRICAL
CONDUIT

Figure 5–7

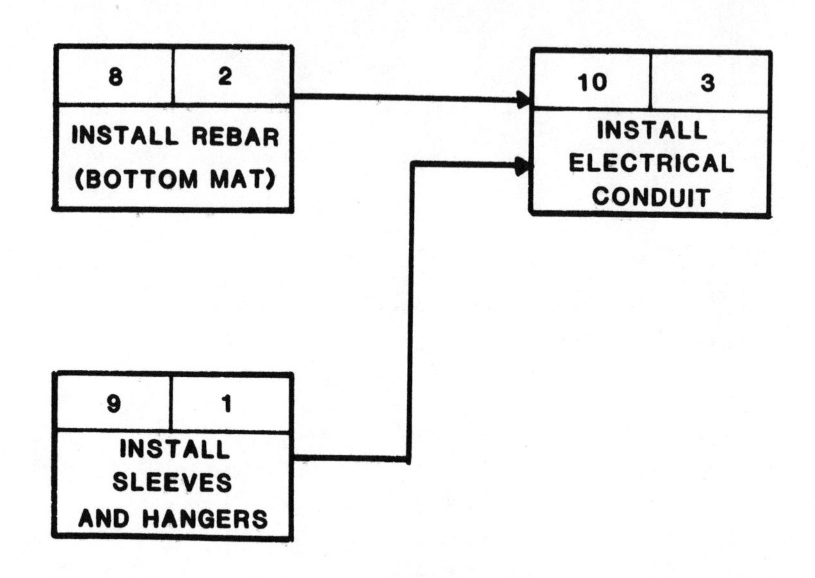
8
2
INSTALL REBAR
(BOTTOM MAT)
10
3
INSTALL
ELECTRICAL
CONDUIT
9
1
INSTALL
SLEEVES
AND HANGERS

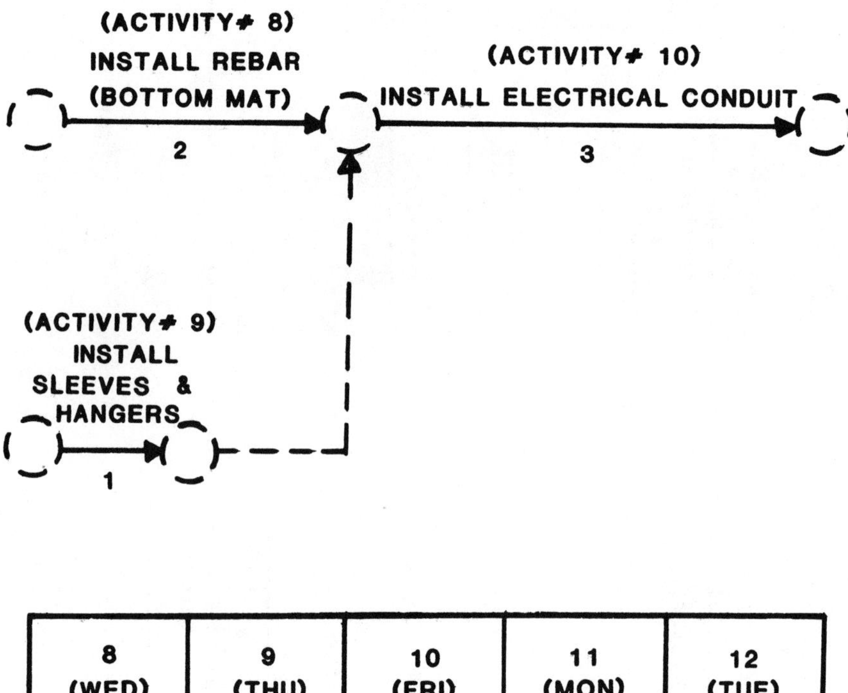
(ACTIVITY# 8)
INSTALL REBAR
(BOTTOM MAT)
2
(ACTIVITY# 10)
INSTALL ELECTRICAL CONDUIT
3
(ACTIVITY# 9)
INSTALL
SLEEVES &
HANGERS
1
8
(WED)
9
(THU)
10
(FRI)
11
(MON)
12
(TUE)

Figure 5–8

Subcontractor's Use of Network Techniques

In addition to understanding the mechanics of network techniques, you, as a subcontractor, should use network schedules in planning and managing projects. Before beginning work on any project, you should review the network which the general contractor is submitting to the owner, evaluate it, and provide necessary input to the general contractor. Under no circumstances should you accept an arbitrary requirement by the general contractor that the subcontract work be performed according to the overall schedule without first taking steps to protect your interests. You should begin by requesting in writing a copy of the overall schedule. This request can be made by utilizing a letter containing the following sample paragraph:

> In order to fulfill our obligations under our subcontract in an efficient and timely manner, we hereby request that you provide us with a copy of the overall project schedule which you will be submitting to the owner. After we have had a chance to review the schedule, we would like to meet with you to discuss necessary or desirable modifications, should there be any.

After receiving the overall project schedule, you should review and evaluate the information keeping several points in mind and offering input to the general contractor when necessary. First, be sure that the general contractor has scheduled the project so that all work which must be done ahead of yours is so scheduled. Conversely, your activities must be scheduled to be completed before logical follow-on work is scheduled to begin. You should be sure that work which is scheduled to be performed concurrently with your activities will not interfere with your forces by preventing access to work areas or causing other inefficiencies. You should be sure that reasonable start dates and durations are provided for your work activities. You must ensure that your intended use and reuse of resources such as manpower (crews) and equipment throughout the project is clearly indicated and evident within your portion of the schedule. Failure to include this information could result in presenting the false impression that it was your intention from the beginning to have unlimited manpower and equipment available to work in virtually all project areas concurrently in order to accomplish your work on the project.

Finally, you should be wary of attempts by the general contractor to reserve float time to its own activities and deny float to you, which could cause you to unrealistically constrain your activity periods. Such a tactic might be used by a general contractor who wishes to set up you and other subcontractors as scapegoats in the event that a delay claim arises.

Use of the Project Schedule during Construction

After you have evaluated the schedule with the above factors in mind, you should provide whatever input you feel is necessary or desirable to the general contractor in order to make the schedule more realistic or efficient. When the performance periods are finally determined, you should use them to develop manloading and equipment loading for the project. Material and equipment purchases and deliveries should also be planned according to the agreed upon schedule.

In developing the planned manpower and equipment loading for the project, you may find it useful to extract your activities from the overall project schedule and to plot them in a time-scaled, calendarized bar chart form. For example, if you were required to follow the "As-Planned" CPM Schedule discussed above, the activities could be represented by two separate bars occurring from day 6 through day 7 and from day 10 through day 12. Thus, the manloading would consist of one crew appearing to perform layout work during the duration of the first activity on the bar graph, and another crew appearing to install conduit during the duration of the second activity. Once you have developed this planned manpower and equipment loading, this data can be used to achieve cost control and to make cash flow projections.

No matter how logical the original schedule is, and no matter how much input you are allowed to provide, the plan will not be useful unless it is updated regularly. Updating your schedule should be part of the ongoing process of updating the overall schedule. In addition to showing percentages of work activities completed, schedule updates should reflect any change in the sequence of activities, any delays, and any accelerations in planned or actual performance. An important part of this updating process is adjustment of the schedule for changes to the work.

An example of an update to a CPM that is based upon as-built data is provided in the "As-Built" CPM Schedule. It may be assumed that this update was issued on day 12 of the project schedule. By that time, two delays had occurred on the job. The first was an electrical strike which delayed the beginning of electrical layout work from the sixth workday to the eighth workday. Meanwhile, the owner and architect delayed another chain of activities by revising the rebar design which added nine days to the planned one day duration for approval of shop drawings. Because of this delay, the electrical subcontractor would not be able to begin installing conduit on the tenth workday as planned. Instead, that work would begin on the seventeenth workday. Based on this information, the electrical subcontractor's manloading plan as it existed when the schedule was updated would have to be revised accordingly.

The final consideration in the use of a network technique during construction involves the possibility of a claim arising on the project. Careful initial planning, based upon the schedule, and consistent and accurate updating during construction help claim analysis in two ways. First, they improve the quality of the claim analysis by allowing you to support your claim position by complete and accurate project documentation. Second, using the schedule in this manner will help reduce the cost of performing a scheduling analysis, since much of the analysis is being done as the job progresses. Construction claims are discussed more fully in chapters seven and eight, but the importance of network technique based schedules and updates in analyzing a construction claim should not be underestimated.

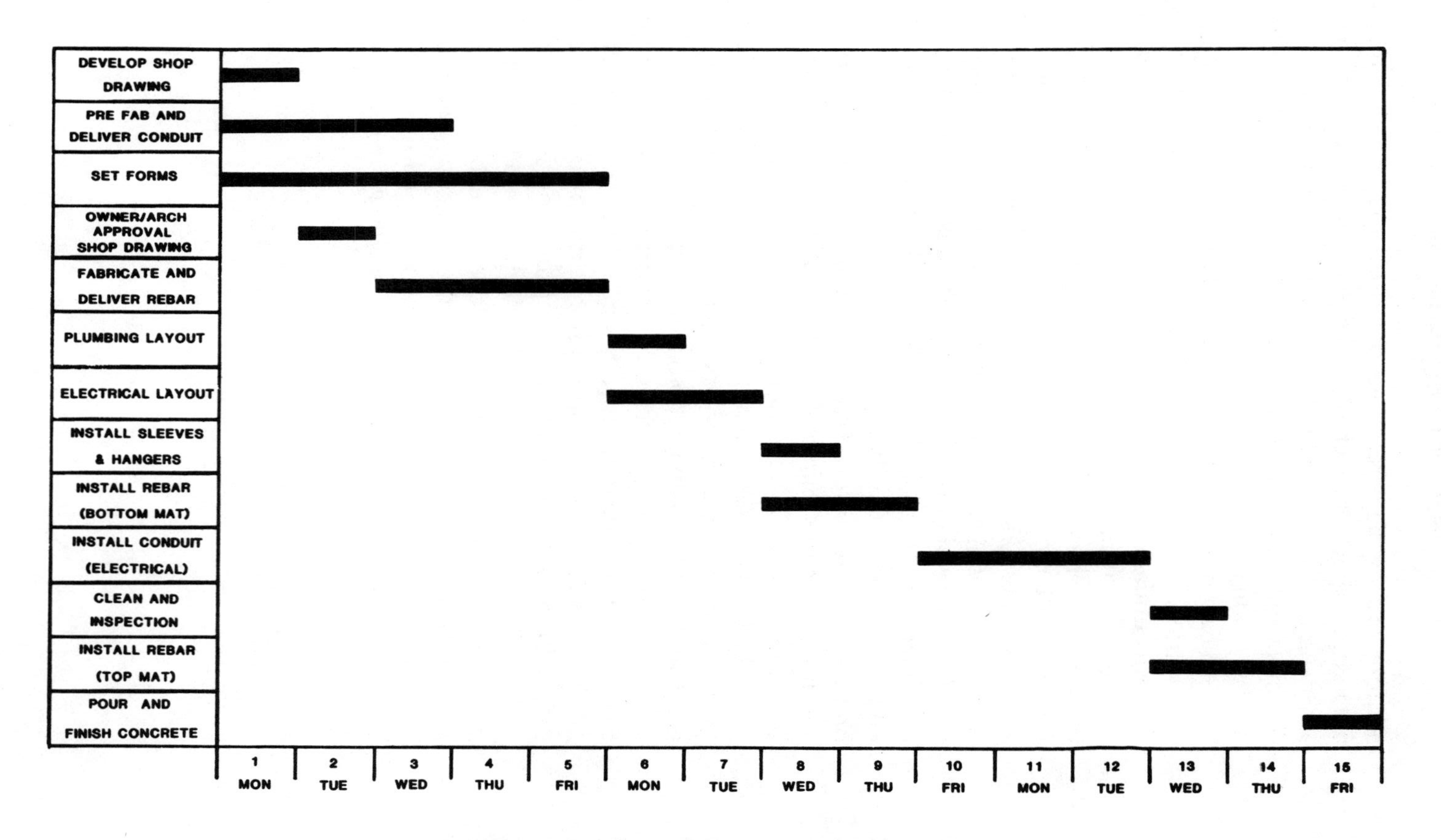

Bar Chart for a Typical Concrete Floor Slab

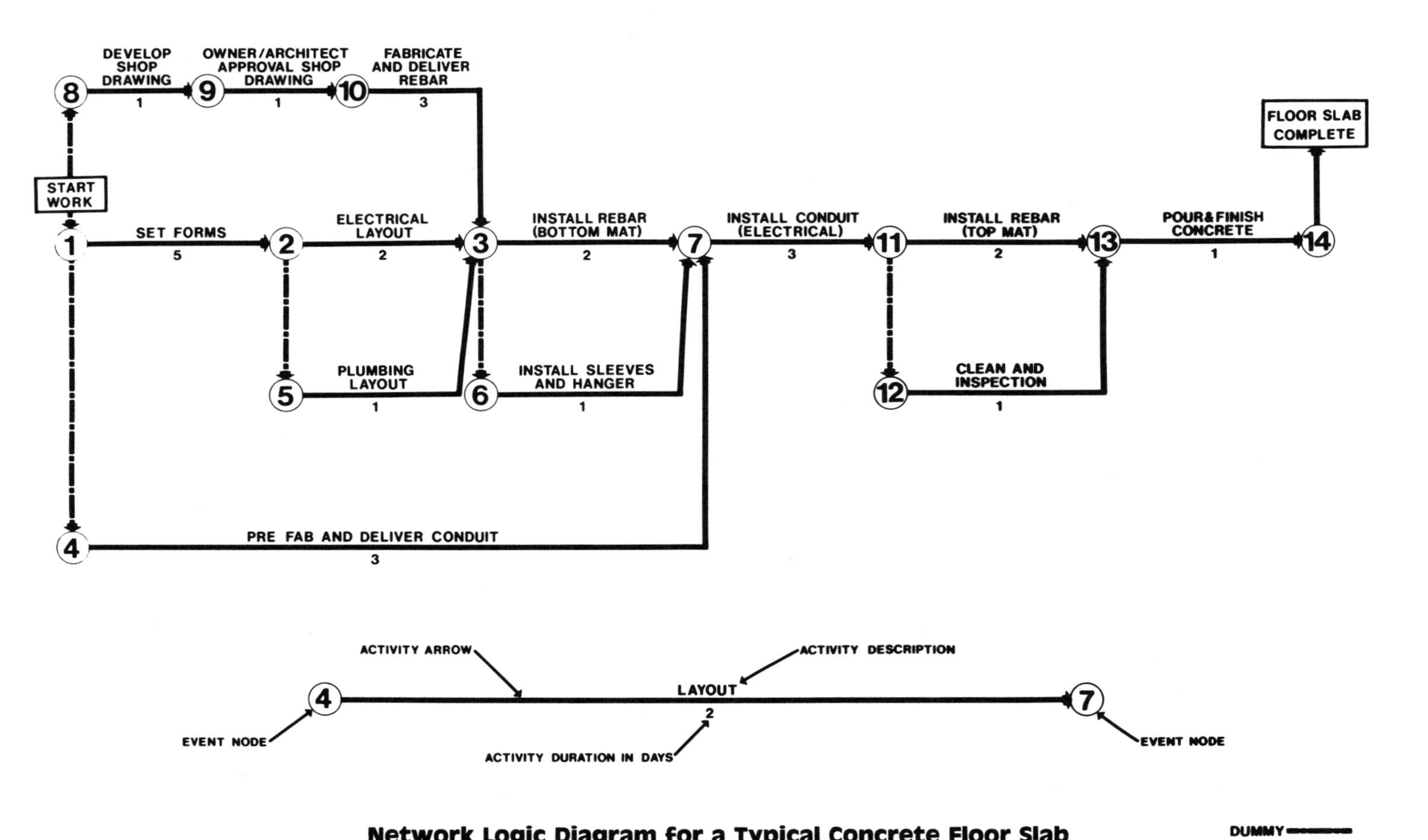

Network Logic Diagram for a Typical Concrete Floor Slab

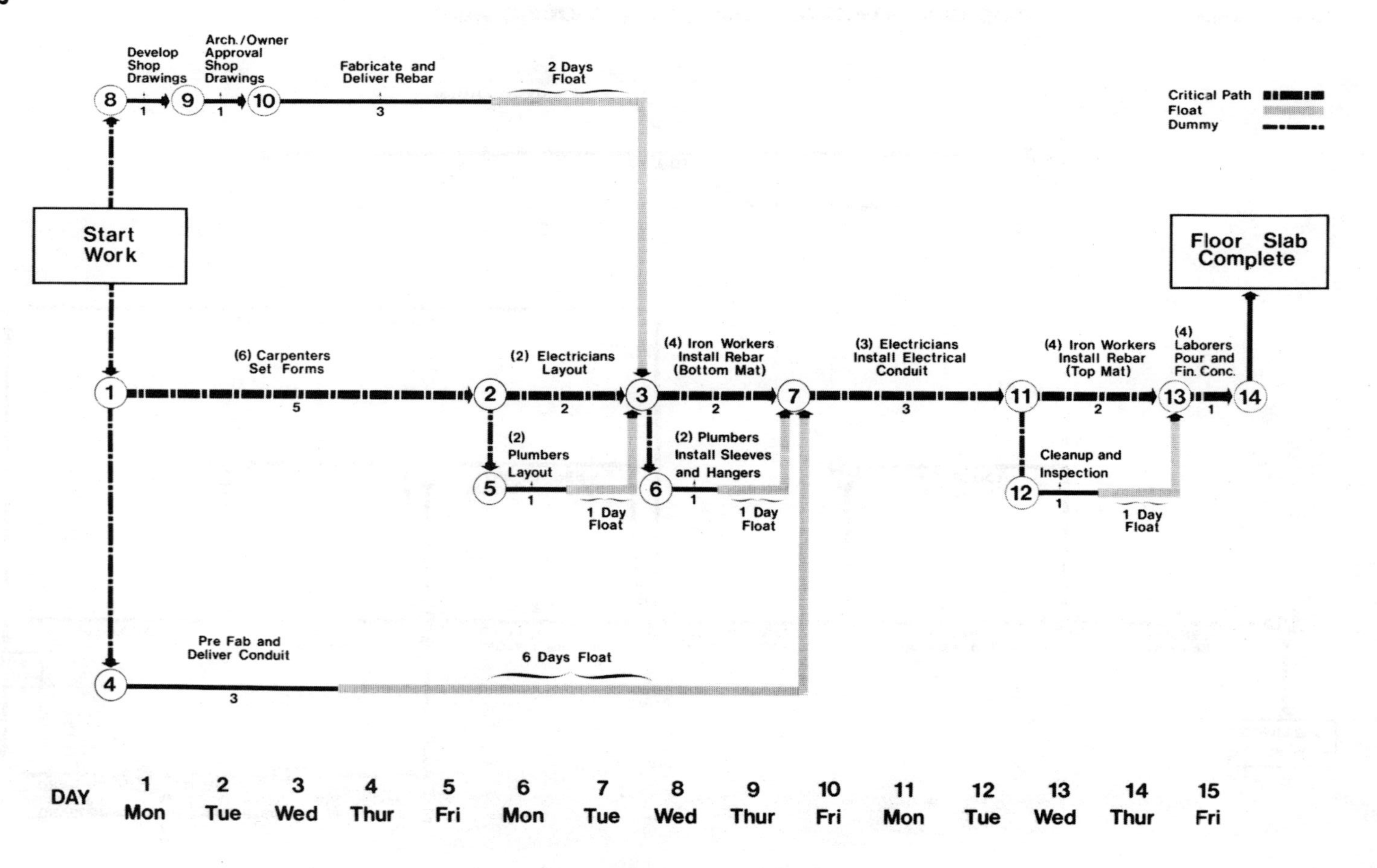

"As Planned" CPM Schedule

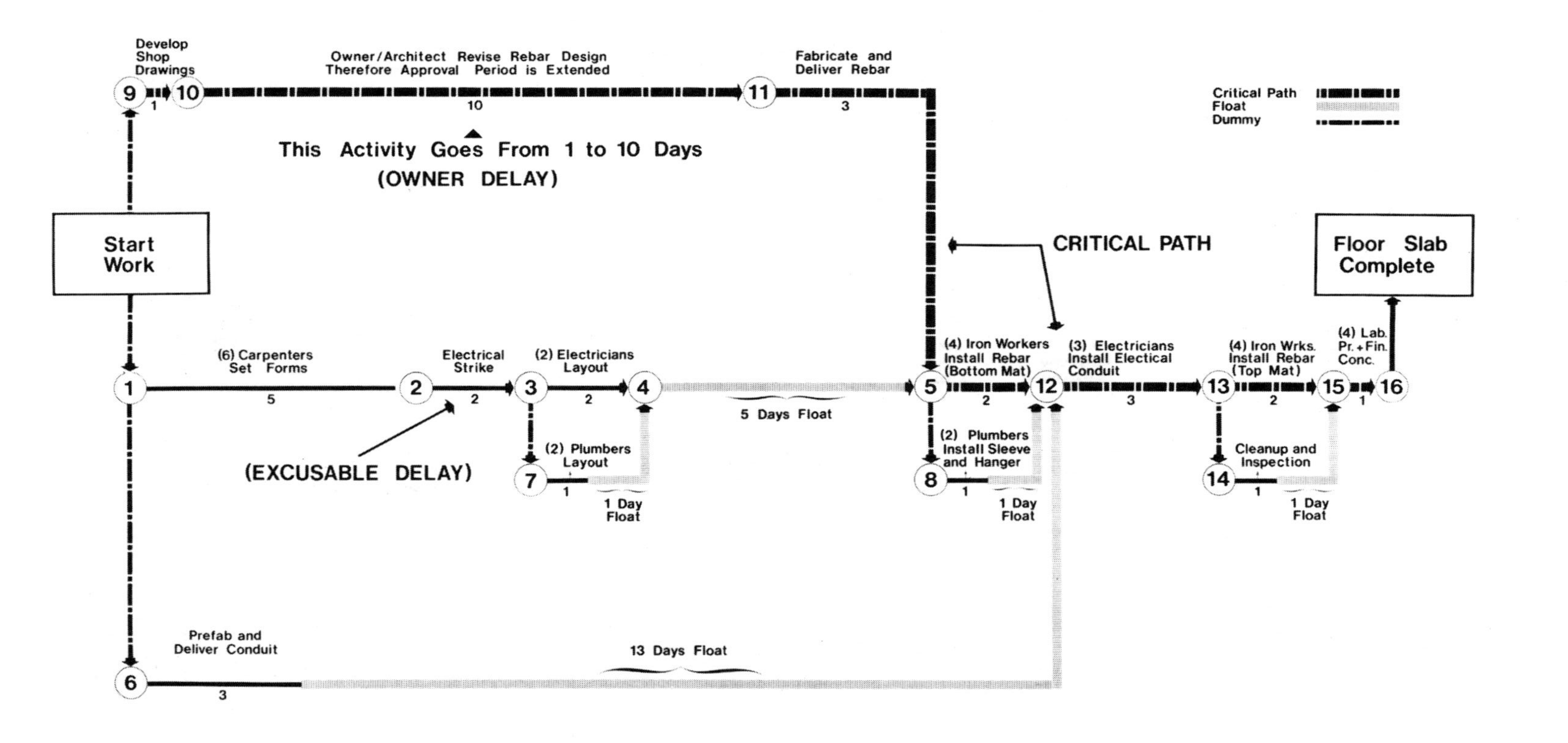

"As Built" CPM Schedule

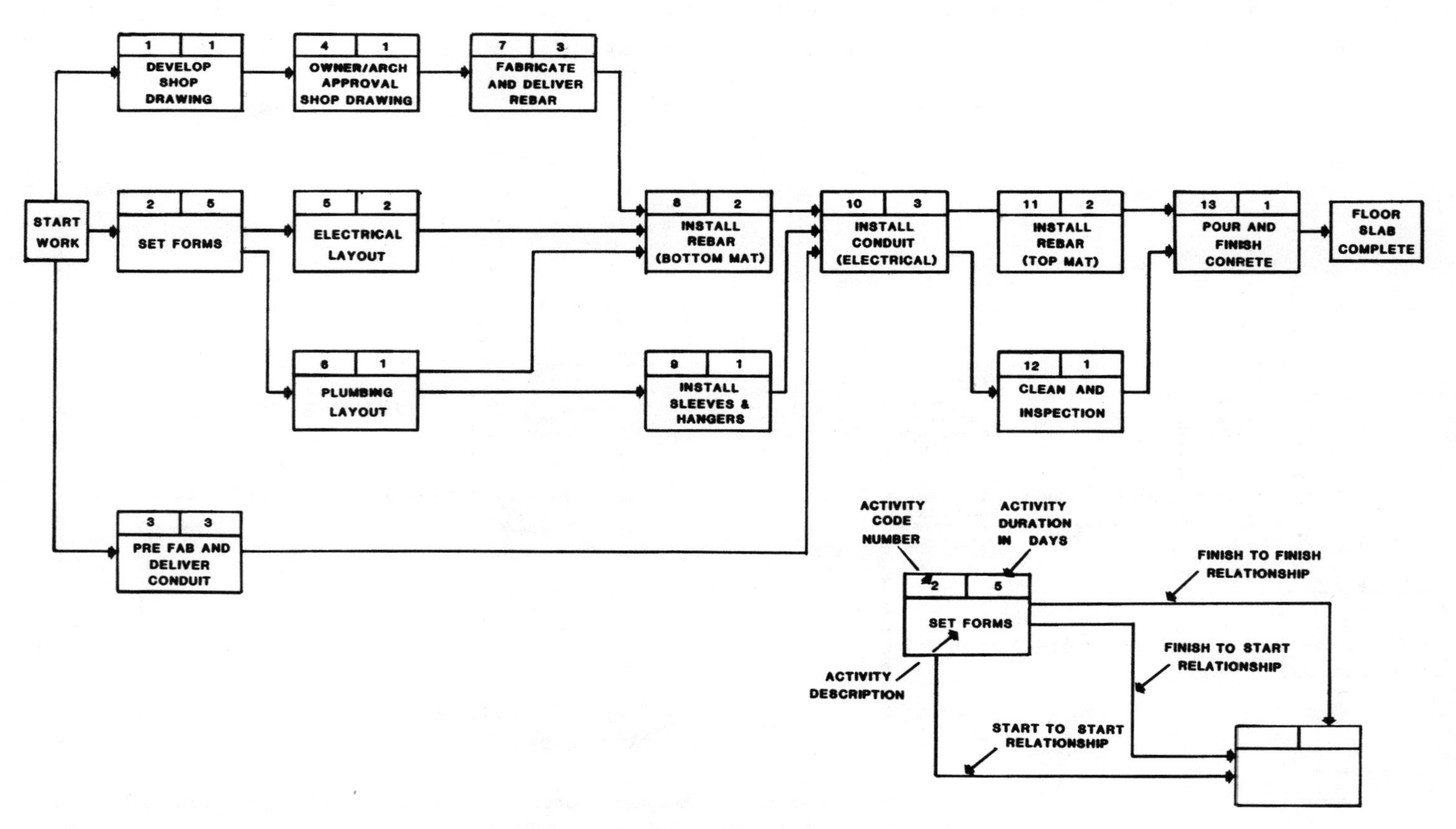

Precedence Diagram for a Typical Concrete Floor Slab

Questions:

1. What are the disadvantages of the bar chart for project scheduling?

2. Why should subcontractors be involved in the preparation of the construction schedule?

3. Why is it important to update the construction schedule?

4. What is the definition of the "critical path"?

CHAPTER SIX

Job Mobilization and Setup

Introduction

After receipt of the contract, you, as the electrical contractor, have a number of key activities to perform which are essential to getting any construction project under way and necessary to form the basis for a controlled, coordinated field operation effort. Regardless of the kind of contract form or the type of construction project, the contractor ordinarily has several detailed requirements in the contract specifications that require that a substantial number of activities be accomplished within the first 60 days of the construction time. Failure to timely complete these activities may result in your waiving rights under the contract to recover compensation for various future actions of the customer. For example, your failure to submit an initial schedule required by the contract may be used by the customer to delay invoice of initial progress payments to your firm.

Therefore, it is of critical importance that you review the contract documents carefully and make a list of the activities that must be accomplished in the first 60 days of contract performance. In addition to these written contract requirements, you should organize your internal project personnel and support forces to establish the proper procedures to monitor and control the construction project costs and performance time. This should be done to enable you to meet your contractual commitments and achieve projected profits.

Since these initial contract and internal organization requirements may be numerous, detailed, and of long term effect, you should have a systematic approach to job mobilization and setup which forms the overall basis for management and cost control of all projects. This approach needs to be sufficiently flexible, however, to accommodate the many unique features of each individual project.

The activities discussed herein should not be construed as all-inclusive. These activities encompass most of the major items that must typically be addressed on a construction project. Some of these tasks will require more detail and attention than others; however, the earlier and the more comprehensively you address these work items, the more likely that there will be an efficient, coordinated and profitable construction effort.

Steps in Effective Job Mobilization and Setup

REVIEW OF CONTRACT DOCUMENTS

Although the key project management team members may have varying educational or legal backgrounds, they must all read and thoroughly understand all of the key contract documents which affect the performance of the work. These documents generally include the agreement between you, the electrical contractor, and the customer. General conditions, special conditions, supplementary general conditions and the contract plans and specifications are included. It is also becoming normal procedure for the general contractor to bind the subcontractor to the terms and conditions of the agreement between the owner and the contractor. If such provisions are included in your contract, you should attempt to obtain a copy of the agreement between the owner and the contractor for your review. If this contract review raises any doubt concerning the intent or meaning of the documents as they are written, you should consider allowing your counsel to review this information in order to obtain an interpretation of the documents. Although the contract documents are generally intended to be complementary, overlaps or gaps often exist in these materials and should be identified as completely and as quickly as possible in order to avoid any delay or disruption in the timely performance of the work.

In reviewing so-called "form contracts," you should keep in mind, during the mobilization phase of your performance, that these "standard documents" are being constantly modified and that supplemental conditions ordinarily included in specifications may modify these standard documents with language that is unfavorable to you as an electrical contractor. Therefore, you should read the supplementary conditions of your contract with particular care.

In order to avoid entering into a contract with unfavorable terms, you may choose to utilize standard form contract agreements or purchase orders, such as AIA Document A–401, or perhaps a form developed by you or your counsel and employed in the past. By drafting the initial contract language, your company has the advantage and opportunity to incorporate the terms and conditions you desire and to utilize your form agreement as a basis for contract negotiation.

As part of your review of the contract documents, key project personnel should also be thoroughly familiar with the individual contract provisions that can impact the progress, and therefore the cost, of contract performance. For example, the project manager, project engineer, and field superintendents should review and understand language in the contract relating to such important items as:

1. The scope of the work.
2. The information or services to be furnished by the owner, general contractor and/or his architect/engineer.
3. The architect/engineer's duties during construction.
4. The owner's duties during construction.
5. The general contractor's duties during construction.
6. Procedures for change orders.

7. The definition of differing site conditions.
8. The subcontractor's right to stop work.
9. Notice requirements.
10. Liquidated damages provisions.
11. Excusable delay and/or time extensions.
12. Payment and completion (failure of payment/substantial payment/final payment)
13. Termination (partial/convenience/default).

If your agreement is on a cost reimbursable basis, the project management team must also thoroughly understand the differences between reimbursable and nonreimbursable costs.

CONTRACT EXECUTION

Although a letter of intent or verbal notice to proceed indicates that you have an agreement, a written notice to proceed obligating the customer to reimburse you for all costs, overhead, and profit during the time of contract performance gives you a much stronger legal and factual position if a dispute concerning the contract later arises. This form of written notice is particularly important since the contract documents may be vague or ambiguous concerning the actual effective date of the notice to proceed. Similarly, you should be wary of entering into an agreement that provides for a notice to proceed to be issued at some future date without including a time limitation on how long this notice to proceed can be withheld. Although this is a matter of business negotiation and judgment, delays in the issuance of the formal notice to proceed may result in escalation rates from your suppliers, as well as in the distortion of your cost basis for the project budget.

PROJECT ORGANIZATION AND PERSONNEL

After reviewing the contract documents, including the scope of work required to be performed, the duration of the project, and the budget categories for supervision, you should establish an organization for the project with clearly defined lines of authority and communication. This decision is not particularly significant on smaller projects which are within the normal operating range of your company's business, since project management personnel normally understand their relative roles and assignments consistent with your company's standard operations and procedures.

The performance of larger projects, however, may require the establishment of an independent organizational structure designed to accommodate the unique conditions of the project, including such items as location and particular requirements of the customer. For example, a decision to locate top project management in the home office or at the job site may be based upon an understanding of the location of the customer's top project management. Similarly, the position, title, authority, and lines of reporting for you and the customer should parallel each other to the extent that a firm understanding is established between the parties as to the proper levels for communications and directions during the course of the project.

As a concomitant element of structuring the project organization, your company must also assign capable and experienced management personnel in order to carry out your obligations on the project. Although the relative duties and responsibilities of the project management team were discussed in detail in chapter two, there are several general criteria which should be utilized in determining the particular staff of a project. For example, many electrical contractors determine staffing according to some of the following criteria:

1. Experience in the type of project being constructed;
2. Experience in the particular area of the project;
3. Experience with the general contractor, owner or architect/engineer;
4. Availability for the duraton of the project; and
5. Willingness to relocate.

BONDS AND INSURANCE CERTIFICATES

As part of job mobilization, you may be required by the contract specifications to provide certificates of insurance and performance and payment bonds. Although their execution and delivery will ordinarily be a routine administrative matter, you should carefully read the indemnity provisions. Many contracts today require extensive indemnification of the parties to the contract, as well as hold harmless agreements that can be extraordinary to your basic insurance coverages and costly to procure. Although the owner may furnish the builder's risk insurance to the general contractor, you should request a copy of the owner's policy and read it carefully to ensure that it inures to the benefit of your company.

EQUAL EMPLOYMENT OPPORTUNITY (EEO) AND AFFIRMATIVE ACTION PROGRAMS

Performance of work on public projects may involve the question of your required compliance with EEO and Affirmative Action Programs. Although commitments concerning minority labor requirements exist, you should check your contract to determine whether these requirements are mandatory or discretionary. In order to determine and carry out your obligations, an early meeting with the compliance agency is important.

SECURE PERMITS

It is common practice that the actual building permit is secured by the general contractor, with the specialty subcontractors securing their own specialty permits. As part of your permit requirements, it is important to obtain information from the general contractor and architect/engineer as to their previous dealings with the authorities. If the authorities have been properly briefed during the design phase of the project, permits are relatively easy to obtain. Therefore, in order to avoid delays in the issuance of permits with resultant impact on the progress of your work, you should endeavor to find out from the general contractor if there are any disputes existing with permitting authorities. In reviewing your contract terms and conditions, be careful not to obligate yourself to obtain permits which are beyond your control, such as zoning or environmental planning permits.

MONTHLY REQUISITION FORMAT AND PROCEDURES

The general conditions of your contract may provide that you furnish the customer with a comprehensive breakdown of all cost items on the project before you can submit your initial progress payment requisition. This breakdown must ordinarily be submitted for the approval of the customer and must conform to his reasonable revisions before he will process the initial progress payment requisition.

It should also be noted that the scheduling provisions of some contracts may require that the monthly payment requisition be compatible with cost items indicated on the approved construction schedule. For example, actual cost loading and the CPM schedule may be the only basis for processing a monthly progress payment requisition.

Progress payment requisitions can take the form of percentages of completion for line items on a fixed price contract, or can be the actual cost incurred on a cost reimbursable project. Regardless of the format for the monthly requisitions, this format and the procedure for submission of the progress payment requisition must be set up during the initial 60 days of the job and should be coordinated with the representative of the customer. Since most contract agreements establish specific dates for submitting invoices and specify turnaround time for processing progress payment requisitions, your field organization should also notify vendors of cutoff dates for submission of invoices so that requisitions can be handled in an expeditious and timely manner. At the same time, you should be sure to include lien waiver and release form information from vendors as part of your progress payment application.

OBTAIN JOB SITE OFFICE SPACE

One of the key job mobilization activities is the design and setup of the job site facilities including trailers, temporary utilities, and construction equipment. If the project requires significant on-site personnel, you should obtain a trailer and ensure that its design incorporates sufficient working room for the required project staff along with necessary conference and meeting areas. To ensure privacy, the office of the project manager or superintendent should be isolated from the general office area.

If the job site is in a remote area, you may encounter problems concerning temporary utility services. You should therefore contact the customer as soon as possible in order to identify the availability of services such as power, heat, water and telecommunications to ensure the hookup of these facilities prior to the commencement of on-site construction.

As part of the setup of your field office, you should consider the needs and requirements of office equipment: typing and reproduction machines, more sophisticated engineering and drafting equipment, telecopiers, and remote telecommunications equipment. To a great extent, the equipment to be installed in your field office will be determined by both the scope of the project and the distance from the job site to your home office. As part of this analysis, you should also consider the relative merits of rental or purchase of equipment. This decision will be based, to a large extent, upon existing equipment resources, as well as decisions made during the estimating and acquisition phase of the project.

PROJECT SCHEDULE

Although your company may have prepared a preliminary schedule during the acquisition phase of the project, you also have the responsibility to develop a more detailed schedule in order to prosecute and manage your work on the project. Although the form of schedule (critical path, precedence method, etc.) is ordinarily addressed in your contract, you should be sure to meet with the customer in order to determine the requirements for your particular schedule.

As was discussed more fully in chapter five, the project schedule can be an essential element for project management. In particular, use of a network scheduling technique such as the critical path method can prove an invaluable tool in monitoring job progress and in early recognition of problem areas for resolution by your project management staff. The proper review and revision of this schedule can also become the basis of any later claim concerning the impact on your progress of changes, differing site conditions or other problems.

For purposes of project mobilization, however, it is important to remember that the planning of the project and the assembling of the basic sequences and durations of your activities should be the responsibility of the project management in charge of supervising the actual performance of the work. Therefore, you should be reluctant to accept, without review, schedules, sequences or durations furnished by the customer. Rather, you should ensure that these parties have the benefit of your input concerning the method and manner of your contract performance.

Having properly mobilized the manpower, equipment, and other resources necessary to perform your work, you will be prepared for an efficient and coordinated construction effort during the field operations phase of the contract. This orderly construction effort, however, will be dependent to a great extent upon the ability and skill of your project management staff.

Questions:

1. What alternatives to using the customer's contract should be considered?

2. What factors affect the organization of the project management team?

3. What responsibility does the subcontractor have for obtaining building permits?

CHAPTER SEVEN

Recognition and Preparation of Claims

Recognition of the Claim

The earlier that a party to a construction contract can recognize the existence of a claim, the better that party can prepare, present, or defend against that claim. For example, when an electrical contractor is able to present the customer with a well analyzed and documented claim presentation, the customer may well grant the corporation additional money or time extensions which it otherwise might reasonably dispute or deny. Such a reaction could be avoided if the customer, like the contractor, had recognized that a claim had arisen at an early date. The possibility that a claim will arise can be recognized by being aware of both the causes and the effects of potential claim situations. For specific areas which should be watched and checked during construction, follow the Claims Avoidance Checklist.

CLAIMS AVOIDANCE AND MANAGEMENT CHECKLIST

Introduction

This guide is provided to you for the purpose of giving a basic understanding of what you should do in order to avoid or manage a claim situation. The information given will also enable you to support or defend a claim, in the situation where one exists.

One area which is extremely important is claims awareness. Your personnel should understand the importance of "being aware" and keeping complete and accurate records of potential claim events during the construction project. Without detailed and complete documentation, the progress of the job cannot be completely determined and the claims, which result from delay, disruption, etc., cannot be verified and supported.

We have prepared this guide to "walk" you through the construction project, with the contract being the first topic. Our guide discusses documentation, scheduling, construction progress, and claims. If you implement some or all of the following suggestions, we do not guarantee you will never be involved in a claim, arbitration or litigation, but you will be provided with the tools to avoid, manage, support, or defend your claim.

Contract

A. Be familiar with the contract documents. Make sure you fully understand the terms and conditions of the contract. *Always* maintain a copy so you can refer to it throughout the duration of the project.

B. Make a checklist of all areas in the contract which require that you give notice and post it conspicuously for your project manager or superintendent to follow.

C. Try to utilize standard form purchase orders and subcontracts for your suppliers and subcontractors in order to shift responsibility to them and likewise limit your liability to them for ownership-caused delays, changes, etc., to the amount which you recover from the owner. Be sure these documents incorporate the terms of the general contract by reference and that all required clauses are included therein.

Project Schedule

A. Determine what scheduling method (e.g., CPM, PERT) is contractually required and in what form it is to be presented.

B. Does the contractor have any alternatives regarding the method and form of scheduling, and is this item negotiable?

C. Are there any questions or problems concerning the method, form or details required relating to scheduling? If so, document and notify the owner of the problems.

D. Make sure that you as the contractor are fully able to supply all the detail and data that is contractually required for the schedule within the time specified.

E. Are the progress payments based upon the schedule, and if so, is front-end loading permitted?

F. What responsibility does the customer have in the preparation of the contractor's schedule?

G. Is approval by the customer or his agent required for the schedule?

H. Is there a contract requirement that the schedule be used to implement or execute the work?

I. Are supplementary schedules permitted at the contractor's option?

J. Are your subcontractors required to receive the schedule and updates?

K. Are there any other contract requirements regarding coordination, such as weekly job meetings?

L. Can the contractor use outside scheduling consultants or does the contract require that someone on the staff have expertise in scheduling?

M. What exculpatory language (e.g., no damages for delay clause), if any, does your contract contain?

N. When are time extensions given and for what reasons?

O. Who gets the benefit of float time?

P. If it is a contract-prepared schedule, does it have to make allowances for holidays, possible strikes, inclement weather, etc.?

Q. Concerning your subcontractors:

(1) Are subcontractors required to provide input for the development of the progress schedule? (Realize this may be viewed as an express warranty in the subcontract that the progress schedule be used by the general contractor.)

(2) Can the subcontractor use the approved schedule or should he be expressly required to follow whatever schedule is issued?

(3) Does the contract specifically provide that you do not guarantee or warrant the duration time for the various items of work or their start and finish dates as set forth in the schedule? You might consider sending the following letter to subcontractors or to the customer when the CPM schedule is forwarded to them in accordance with the contract requirement:

" Re: CPM Schedule
Contract No. ________

Dear ______________________________ :

Enclosed please find our proposed CPM schedule for the project.

Except for contractual completion requirements, the durations given for the various items of work shown on the enclosed schedule are not commitments. They are estimates based on conditions currently known or represented and any changes discovered will result in a change in the schedule. Please note specifically those items of work which are related to or dependent upon actions under your control, since we rely on being able to commence and proceed with those activities on the dates shown and any delays of that nature will affect this schedule.

We assume that this schedule is acceptable to you and in conformance with our contract requirements unless we hear from you to the contrary within the next ____ days.

Very truly yours,"

Documentation

Documentation is one of the most essential elements in construction today. Record-keeping systems should be complete and should include the following documentation:

A. The contract (including general, supplementary and special conditions);

B. Progress payment requisitions and all data which was submitted to support these requests;

C. Inspection reports;

D. Daily logs and journals maintained by on-site personnel or project representatives. Journal and daily log entries should relate *only* to the events and progress of the project;

E. Daily reports which include the area in which work was performed on that day, the time spent by each person, and any indication of problems and delays occurring on the project;

F. Memoranda to the project file which relate to conversations, inspections, directions, observations, and any delays or problems.

G. Correspondence. *All* correspondence should be maintained in one central file;

H. Photographs which should be taken on a regular basis to illustrate construction progress and also, whenever necessary, to document a particular problem which has arisen in the field which may delay, disrupt, or interfere with the progress of that work;

I. Minutes of weekly job meetings or other meetings held between the contractor, the customer or personnel. All minutes of meetings should be sent to other persons present at the meeting for their comments;

J. All cost reports of any nature maintained on the project;

K. Copies of approved or rejected shop drawings and shop drawing logs as well as other submittals;

L. Internal interoffice correspondence;

M. Any analyses made during the performance of the project concerning labor productivity, scheduling analyses, activity duration analyses, or any other related subjects done by in-house personnel or outside consultants;

N. All schedules should be maintained. This includes the initial schedule submitted to the customer or subcontractors, the initially approved schedule, any marked-up schedules used in the field for updating, any supplementary schedules (i.e., pour schedules, finish schedules or detail daily or weekly schedules), any in-house schedules prepared for subcontractors or suppliers, and any hand-drawn schedules which were done on an informal basis in the field to explain the mechanics of a particular trade. Make sure to always obtain and record all important information in writing, and make sure that it was *signed* and *dated.*

Job Progress

A. It is advisable to hold regular job progress meetings to discuss any problems or delays which may occur.

B. Analyze your job progress in detail prior to the progress meetings and use the meetings to discuss new delays and problems. Also, inform your personnel of current events and job progress.

C. Request time extensions whenever significant events occur which may entitle you to same under the contract.

D. Record all conflicts, discrepancies, etc. Inform the customer as delays occur so that you will be properly coordinated and not incur responsibility due to lack of information.

E. Obtain written confirmation of all oral directives. If the party will not confirm in writing, you should write a confirmation letter to them.

F. Maintain a log of all drawings, submittals, change orders and proposals, pertinent correspondence, and other similar data for ease of reference and also to indicate changes.

G. Review employees' time cards or Foreman's Daily Reports in detail and revise their format as necessary to better reflect potential claim areas such as extra work being performed, delays, etc.

H. Cost code your projects, i.e., separate your costs into areas and types of work. This method will enable you to better analyze a potential claim in the areas of costing. This cost code should be established in the initial stages and updated throughout the project.

Claims

In the event of a claim situation, the contractor should keep all areas open by writing and filing *written* notices of potential claims to the customer or his agent within the time specified in the contract after such time as the contractor feels they have been delayed, disrupted or accelerated. The contractor should recognize a potential claim when one or more of the following conditions occurs:

A. Extra work not specifically provided for in the drawings and/or specifications;

B. Work in a particular manner or by a particular method which varies from or is more expensive than the method originally anticipated;

C. Work according to specifications or drawings which have been changed, amended, revised, amplified or clarified;

D. Work according to drawings which are so limited in detail as to require the contractor to perform unanticipated engineering services;

E. Work according to one particular method when two or more alternative methods are allowed by the contract, or when the contractor should be free to devise his own methods;

F. Work performed out of the planned or normal sequence;

G. Stop, disrupt, or interrupt work to any extent, both directly and indirectly;

H. Work in congested work areas or lack of available work areas;

I. Owner or other prime contractor-furnished equipment delivered late, in poor condition or not suitable for the use intended;

J. Accelerate performance in any way to regain schedule, to add men, equipment, or materials, or to work overtime or extra shifts;

K. Compression of work; follow any schedule which varies from your approved schedule or planned sequence of work;

L. Relocate existing work because of lack of coordination or information.

Do not waive claim rights by negotiating or signing off on contract modifications or change orders which only pay part of the excess cost (i.e., usually only direct costs), and those which do not provide time extensions or impact costs as needed. This is a critical item in that the disruption costs, extended overhead, wage rate, material escalations, acceleration costs, and other impact costs are not always apparent at the time a change takes place. In such cases, it is recommended that the contractor insert a reservation of rights in the change order proposal or change order itself.

CAUSES OF CLAIMS

The cause of a construction claim is always particular to that claim, but certain recurring general circumstances should alert the parties to any construction contract of a claim situation. Some circumstances which may result in claims by or against your corporation include the following:

1. The issuance of an extensive amount of extra work orders or changes to the original contract's scope of work.
2. The existence of site conditions which differ from those described in the contract documents or the information provided to bidders.
3. Inadequacies or inaccuracies in contract drawings or shop drawings.
4. Interference by the customer or other parties to the project with the contractor's manner and method of performance, including failing to provide access to the job site, requiring the contractor to perform according to one method when the contract documents provide alternative methods, issuing directives to the contractor's forces directly, and issuing orders to stop, or conversely, to accelerate work on the project.
5. The installation of work which fails to conform to the contract requirements.
6. Lack of coordination and cooperation between forces of the subcontractors and of the prime contractor.
7. Lack of adequate supervision and manpower at the job site.

Many of these various causes of claims are exemplified by reported decisions of courts or boards of contract appeals concerning disputes arising out of work performed by electrical contractors. For example, an electrical contractor successfully prosecuted a delay claim based on lack of site access at a federal project in Chicago, Illinois, on which it was one of several prime contractors. The contractor's site access was delayed by an earth movement which occurred about three months before planned commencement of work at the site. This earth movement damaged piles and delayed foundation and structural steel work causing the electrical contractor to suspend its planned start of work at the site for 322 days. As a result of a hearing before the GSA Board of Contract Appeals, it was determined that the contractor was entitled to compensation for costs incurred during that period according to the standard Suspension of Work clause contained in its contract.

A case involving the construction of a postal facility provides an example of a claim arising out of a requirement that the contractor perform according to one method when the contract documents provide alternative methods. The specifications in question stated that the contractor was to install photoelectric cells "within or adjacent to" mailing vestibule doors through which postal employees would be driving carts. The contractor submitted shop drawings showing the photoelectric cells on the door jamb, i.e., "within" the doors. The architect/engineer disapproved this arrangement and directed the contractor to locate the cells on vertical steel tubes just beyond the point to which the doors would swing when opened. This installation required the use of additional conduit and wire. Since the contractor was denied the opportunity to install the cells in accordance with the choice he was entitled to make under the contract documents, he was entitled to an equitable adjustment for the increased costs incurred in installing extra conduit and wire.

A contractor successfully pursued claims for extra work and acceleration arising out of a project to construct and install radio transmitting and receiving facilities in the Philippines. The contract included the fabrication and erection of 104 steel towers to support the transmitting antennas. Upon receipt of the first delivery of steel, the government questioned the quality of the steel because of seams appearing on the surface of the material. The government immediately stopped work and, after several months of considering solutions to the problem, required the electrical contractor to perform substantial corrective work. Because the steel in its original form met all specifications, the electrical contractor recovered the costs of the extra corrective work. In addition, it was entitled to compensation for the costs of accelerating the job to get back on schedule after the government resolved the problem of the seams.

EFFECTS OF CONSTRUCTION CLAIMS

Since every problem which could potentially cause a construction claim does not in fact do so, parties to construction contracts must also be aware of the effects of a claim situation in order to recognize when one truly arises. The more significant manifestations of a construction claim include the following:

1. Most obviously, notices from and to subcontractors, the prime contractor, and the owner and his agents. In addition to general notice of a claim, notice may be given for lack of progress payments, filing of mechanics' and materialmen's liens, and the issuance of change order proposals.

2. Disparities between the contractor's planned performance and his actual performance. When the job falls behind the planned schedule of performance, all parties to the contract should be alerted to the possibility of a claim. Falling behind schedule is significant at any stage or phase of a project. This includes delays in the issuance of plans, the provision of site access, the issuance of progress payments, and the approval of shop drawings, proposed changes, and other contractor submissions, as well as the basic construction of the project.

3. Even if construction is proceeding ahead of schedule, a claim may also arise if work is being performed in sequences which differ from those originally planned. Such circumstances typically indicate the presence of an inefficiency or lack of productivity claim. A delay claim, however, may also exist if the work could have been performed both faster than it was actually performed and faster than the original schedule indicated.

4. The existence of direct cost overruns. When the actual cost for labor or material greatly exceeds the originally estimated cost, it should indicate one of two things: the contractor un-

derestimated that item, or a claim, probably for delay or disruption, has arisen on the project.

CLAIM ORIENTATION

In order to be successful in a claims situation, it is not sufficient for parties to construction contracts simply to be able to recognize, in a timely manner, the causes and effects of construction claims. Each party to a construction contract must actively monitor its contract performance in order to give itself the opportunity to recognize the claim by discovering evidence of the claim's existence. The responsible parties must continually review their correspondence in order to find notices of claims. They must update their project schedules and compare them with actual job progress throughout the project. Actual costs should be reviewed in the same fashion to discover cost overruns and possible claim areas. It is essential that such cost reviews and scheduling updates be conducted on at least a monthly basis. Many contracts provide that claims must be filed within thirty days or less from the time they occur; otherwise they are waived. A timely review of correspondence, schedules, and cost information gives you a chance to avoid any arguments concerning waivers of claims.

The fact that a party to a construction contract is claim-oriented does not mean that that party will or should file a claim every time it recognizes one. For example, a contractor may have a claim on a job which assures him an overall profit. In addition, the contractor may have a legitimate expectation of receiving future work from that customer. In such circumstances, you, as the contractor, may be well advised for purely business motives to reduce or forego your claim. You cannot, however, be in the position to make that judgment without first being able to recognize and document the existence and amount of the claim.

Parties to Construction Claims

A party having a claim must file it against the proper party on the project involved. Because of the multiple number of parties and contractual relationships involved in a construction project, one claim often perpetuates or fosters another. The following hypothetical case sets forth such a situation and illustrates and identifies the proper claimants and their respective positions.

> The owner hires the prime to build an office building designed by the architect who subcontracted the structural steel design to the engineer. The prime subcontracts the electrical work to the electrical sub and buys the structural steel from the steel supplier. As a result of a failure in the structural steel, all work is suspended for six months while the necessary corrective design and corrective work are done. The failure is in part due to inadequate design and in part due to the supply of inferior quality steel. The question is: Who has a claim against whom?
>
> a. The electrical sub has no breach of contract claim against the steel supplier or the engineer, although they were the parties ultimately causing the structural steel failure, because the electrical sub has no contract with either party. Instead, the electrical sub has a contract delay claim against the prime based upon a suspension of work.
>
> b. The prime also has a delay claim against both the steel supplier and the owner. Both are breach of contract claims, but the claim against the steel supplier is governed by the Uniform Commercial Code since it is based on a contract for the sale of goods. The prime, in addition to having its own claims against the owner and steel supplier, has the right to pass through the claims of the electrical sub under an indemnity theory to each party, the owner bearing the damages due to defective design and the steel supplier bearing the damages due to defective materials.
>
> c. The owner, in turn, has the right to pass through breach of contract claims based on defective design, from the prime to the architect.
>
> d. The architect may then pass on the defective design claim to the party ultimately at fault, the engineer.

e. The steel supplier and the engineer, or their respective insurance companies, must ultimately bear the loss of the damages which they caused.

Claims generally are only filed against those parties with whom the claimant has a contract since claims based upon breach of contract theories require the existence of a binding contract. In recent years, however, courts have been willing to allow claims against parties at fault when there is no existing contract, if the claimant relies on a negligence theory. The clasic example would be a claim by the prime in the above hypothetical against the architect and the engineer on the theory that they negligently designed the structural steel. Although such a negligence claim may allocate fault in the most direct manner, it may also be more difficult to prove the elements of a negligence claim than the elements of a breach of contract claim. This factor coupled with the issue of whether the claim is being pursued through negotiation, arbitration, or litigation, are important considerations in determining against whom a claim will be filed.

Documentation of Construction Claims

IMPORTANT CONTRACT PROVISIONS

Every party to a construction contract should be thoroughly familiar with the provisions of the contract which are relevant to construction claims. These provisions include clauses which state the circumstances under which a party to the contract is or is not entitled to an adjustment in contract price or time extension. These provisions typically include a changes clause, a differing site conditions clause, a suspension of work clause, and a liquidated damages clause. Each of these clauses should be carefully reviewed for any peculiar language setting forth conditions for entitlement to financial compensation and/or time extensions.

In addition, you must carefully follow the language in these provisions and elsewhere in the contract which states requirements of written notice of claims. Your failure to give notice within the contractually prescribed period may result in the waiver of a significant claim. This harsh result does not always occur, however, since courts and boards of contract appeals have developed several exceptions to the rule that written notice provisions will be strictly enforced. For example, a customer may waive written notice of a claim for extra work when he has actual knowledge of the work and is not prejudiced by the lack of written notice. Likewise, a customer may waive the written notice requirement by failing to write extra work orders despite a contract requirement that he do so. Despite the existence of these exceptions, a claimant occupies a stronger position when he gives written notice as required by the contract and thus renders such arguments unnecessary.

RECORD-KEEPING

A wide variety of factual records is developed during the ordinary course of any construction project for purposes of job control and coordination. As explained below, many of these same records are critical to the analysis, presentation and defense of a construction claim. Below is a list of various types of documentation which should be kept and updated, as it will aid you in the presentation or defense of a claim.

1. Original Estimate and Bid Documents;
2. Contractor/Subcontractor Contract Documents;
3. Contractor/Subcontractor Correspondence Exchange;
4. Subcontractor/Supplier/Manufacturer Correspondence Exchange;
5. Superintendent Daily Log/Reports. (Include notes of the following: when activities start or finish; when Owner's contractors start or finish; and interference caused by Owner's representatives or contractors.);
6. Monthly In-House Progress Meeting Minutes;
7. Memos to File/Memos for Record;
8. Original Schedule and Updates;

9. Utility Records/Other Agency Records;
10. Special Owner/Architect/Contractor Meeting Minutes, Notes, Logs, Etc.;
11. Invoices and Purchase Orders;
12. Monthly Progress Payment Requisitions, especially those which show when Owner paid Contractor for an item of work;
13. Change Order Logs;
14. Change Orders and Records of Negotiations;
15. Plans and Specifications;
16. Statements of those who worked on the project and were in a position to witness delays;
17. Photographs (Job Progress and/or Special);
18. Subcontractor's Foreman/Superintendent/Project Manager Logs/Reports;
19. Shop Drawing Logs and Submittal Data;
20. Independent Testing Agency's Reports;
21. Payroll Records;
22. Weather Data;
23. Strike Records;
24. Other Claims by Contractor/Previous Contractor/Follow-on Contractors;
25. Prior/Follow-on Contractors Records/Personnel;
26. Cost Budgets, Unit Cost Reports, Manpower Budgets and Labor Cost Projections;
27. Monthly Subcontractor Meeting Minutes;
28. Monthly Owner Meeting Minutes.

The discussion to follow provides a more extensive explanation of the importance of particular documentation. The preceding list will provide a checklist for you and can be used as a quick reference.

Progress Payment Requisitions

Progress payment requisitions are of central importance where there is a claim based upon the failure to make proper progress payments. They may also be used in analyzing actual construction progress in the form of an as-built schedule, since they show the actual percentage of work completed at regular intervals during project performance.

Daily Diaries

The job superintendent and project manager may maintain personal records of what is happening on the job both as a means of monitoring job progress and as a means of documenting potential claims. Such diaries should record any significant claim-related events, notes of important meetings and telephone conversations, and a description of the general progress of the project.

Daily Reports

The job superintendent should compile and maintain a set of daily reports as prescribed in chapter four. These reports should set forth, in detail, all the information relevant to any claim. Such reports can be used in claims analysis and in presentations to show actual job progress and actual utilization of manpower, labor, and equipment by the contractor and various subcontractors.

External Correspondence

All correspondence related to the project should be maintained in an easily understandable filing system. This correspondence should include memoranda of telephone conversations and letters confirming telephone conversations. By consistently presenting your company's position in correspondence, the project manager involved in a construction claim facilitates the defense or presentation of the claim once it is filed. Correspondence between parties contesting a claim is invaluable in recreating the circumstances which gave rise to the claim and is crucial in proving whether timely notice of the claim was given. To the extent possible, the project manager and his staff should develop form letters for transmittal of shop drawings, daily reports, and other recurring correspondence.

Photographs

Photographs should be taken for two principal purposes. First, they should be taken on a regular basis to show general job progress, regardless of the existence of a claim. General progress photographs can be significant in proving or defending a delay or acceleration claim which, oftentimes, does not arise until a later date. Second, they should be taken in order to illustrate specific job problems. Since those who ultimately evaluate and decide the merits of a claim are often unfamiliar with actual job conditions, photographs can prove to be an invaluable part of a claim presentation. In either circumstance, photographs should be of good quality and labeled by date, location, subject matter, and photographer.

Meeting Minutes

Whenever there is a job progress meeting of any kind, a representative of your company should attend the meeting and record his own minutes. After having done so, he should send a copy of the minutes to all parties who attended the meeting, with a request that they notify you of any inaccuracy. If you should receive minutes from another contractor, you should read them carefully and notify that party of any inaccuracy, whether requested to do so or not.

Cost Records

A complete set of all cost records for the project should be maintained. These records should include all computer printouts, as well as supporting information such as purchase orders and payrolls. In addition, cost records should be set up according to the cost coding system described in chapter three. This allows accurate tracing of cost overruns to particular, inefficient, base contract work, change orders, material escalations, etc.

Drawings

A complete set of all original contract drawings should be maintained in addition to all shop drawings. These drawings should be marked-up as described in chapter four to show actual job progress. A log of shop drawings showing dates of submission, rejection, and approval should be maintained as a means of showing the timeliness of actions on such drawings.

Internal Correspondence

It is very helpful to maintain internal memoranda which will support your company's position should a claim arise. However, your company must be cautious about maintaining internal memoranda which reflect unfavorably upon your own actions, since such memoranda are generally available through discovery to an opposing party. Such memoranda may be covered by the attorney-client privilege and therefore do not have to be produced for discovery.

Schedules

All schedules developed for the project, whether or not actually used in construction, should be maintained as part of the project records. Preliminary schedules may be used during a claim to show how an approved plan was logically developed or how the originally anticipated schedule was adversely affected by problems arising at early stages of the project. The original schedule must be maintained since it is in most cases the principal document upon which any claim is based. Revised schedules should be kept as a means of showing your attempts to mitigate damages by maintaining job progress and by dealing with problems which made the original schedule inadequate.

The Claim Package

Prior to the discussion of the claim package, it should be noted that you can prepare a claim package without the costly advice of your counsel. In the pursuit of construction claims, cost of attorneys is an important consideration. Oftentimes, an attorney is retained prior to the time they are actually needed. In the past, the trend has been to get the attorney involved in the early stages of a claim. This has been a costly, unnecessary undertaking for many contractors. During the last couple of years, construction consultants have been sought to develop claim packages and analyses, since they have the technical expertise in claims disputes, including practical field construction, schedule, and cost analyses experience. A consultant will work with you to help solve the dispute or to assist in the documentation and preparation of a claim. The construction consultant is experienced in claims preparation and can recommend what analyses should be performed in order to best represent your position. After assisting in the preparation of a claim package, the consultant will help you prepare for negotiation, and if negotiation is unsuccessful and formal legal action is required, the construction consultant can prepare all the factual material, charts, and graphs to be used in court or arbitration. It is not uncommon now for an attorney to be retained only at the time of formal litigation (i.e., to file all the appropriate legal documents), instead of when a claim is first discovered. With the expertise of a construction consultant, the contractor does not need to incur the costly fees of an attorney until such time as an attorney's expertise is required to litigate the claim.

FORMAT

A substantively valid claim can be pursued to its fullest extent only if it is presented in a convincing and understandable format. This presentation is usually best accomplished through the utilization of a claim package or claim document which can be prepared with the assistance of a construction claims consultant. The claim package must be easily presentable and understandable. Physically speaking, this rule means that the document must be neatly typed and securely bound, preferably in one volume. Conceptually speaking, the claim package must present an orderly statement of the contractual and factual bases of the claim and of the relief that is being sought through the claim package, whether a contract price modification, time extension, or both. The presentation of the conceptual claim analysis is more fully discussed in the following sections.

Regardless of the logical organization of the claim package, the claim must be based upon exhibits or demonstrative evidence of some kind in order to be credible. Correspondence, photographs, daily reports and the other categories of project factual documentation previously described should be included in the claim package when they provide significant support for the claim. In addition, analyses or graphic illustrations developed from daily reports, pay requisitions, and other sources of documentation should be included in the claim package as a means of summarizing the factual basis of the claim. A construction consultant can help you to develop the appropriate analyses to support and strengthen your claim, but the contractor must have the proper records.

There are two possible methods of handling the inclusion of exhibits in a claim package. First, they may be maintained in a separate appendix to the claim document. Second, to avoid the necessity of constantly referring from the body of the text to the appendix, exhibits may be included within the text itself. This latter method also emphasizes the factual nature of the claim.

INTRODUCTION TO THE CLAIM PACKAGE

The claim package should begin with a short section providing the reader with general background information about the project. You should also keep in mind that the persons evaluating a claim package are not necessarily those persons who are intimately familiar with the project. Therefore, this section of the claim should briefly describe the kind of project and the scope of work set forth in the contract documents. It should identify the principal parties involved in the project and their positions in the claim situation. In addition, the introductory claim section should explain the overall organization of the project. For example, the introduction should explain whether multiple prime contracts or the more traditional single prime contract with multiple subcontracts was used by the customer in setting up the project. The introduction to the claim should conclude by briefly summarizing the principal factual problems and the contractual basis which gave rise to the claim, and the monetary relief or time extensions sought by the claimant.

CONTRACTUAL AND LEGAL BASIS OF THE CLAIM

A claim package is not meant to be a substitute for a legal brief for a court or board of contract appeals. Rather, the primary purpose of a claim package is to provide the claimant with an effective tool for requesting and negotiating a settlement of his claim. Nevertheless, it is useful to set forth the legal basis of a claim in order to impress upon the recipient the claimant's contractual and legal rights to the relief requested.

This section of the claim package should generally cover the contractual and legal basis for entitlement to monetary compensation or time extensions as well as the contractual and legal basis for the method of calculating damages. An explanation of entitlement to relief should always begin with a discussion of the express provisions of the contract, such as the changes clause or a general clause covering equitable adjustments. In support of your claim, it is also useful to cite well-recognized implied contractual duties such as the duty to cooperate, and the duty not to interfere with the performance of the other contracting party, if applicable to your claim situation. These duties, which are implied in all construction contracts, are especially useful in claims arising out of customer interference with your work or failure to provide site access to your company. The discussion of contractual and legal principles in the claim package should also set forth contractual and legal authority for the recovery of the categories of damages being sought since the extent of liability, rather than the existence of liability, is often the most difficult issue to resolve in a construction claim. Finally, any section of a claim package which touches on legal rights and remedies should be reviewed by the claimant's contract administration department and attorney before the package is submitted.

NARRATIVE

The claim package should contain a thorough statement of the significant events which transpired on the project and which contributed to the claim. It should be clear from a reading of this factual statement that the legal and contractual rights and remedies described in the preceding section are applicable to the claim being described.

This is also the portion of the claim package that will contain the documentary and graphic exhibits previously described. Because the claim package may be the reader's first opportunity to comprehend detailed information regarding the claim, it can be more useful to place exhibits throughout the narrative as they are referenced. This approach helps to preclude the impression that the claimant has manufactured his claim after the fact.

The narrative of the claim should have a logical internal organization, for example, chronological or topical. Every claim package is unique in this respect, and the most effective format for a narrative will be suggested by the circumstances of the claim itself. If a claim package is divided into various subclaims, each subclaim should be factually supported so that it may stand alone. This organization allows give and take in negotiations and prevents the denial of all claims because of one weak subclaim.

DAMAGES AND TIME EXTENSIONS

The statement of the relief being requested may be the most important section of the entire claim package. For example, the party defending the claim may be fully aware of contract performance problems such as delayed site access, defective plans, and differing site conditions, for which you are requesting an equitable adjustment. But that party is not necessarily ready to grant whatever relief you request. Since a rational customer is not about to give you anything to which you are not entitled, the contractor should be aware of the types of relief which may be requested depending on the type of claim being presented. Some of the basic claim categories and corresponding categories of relief are described below.

Delay Claims

Most construction contracts for projects of any significance include clauses covering delay in performance and time extensions. These clauses typically provide that for causes of delay which are not foreseeable or not under the control of either party, such as strikes, unusually severe weather, and acts of God, you are entitled to an appropriate extension of time for contract performance for the resulting delay. But you are not entitled to any related additional costs incurred. Thus, these delays are termed noncompensable.

Construction contracts generally provide no relief to you, as the contractor, when you are responsible for the delay, and at least where liquidated damages for delay in performance are stipulated, expressly require you to pay damages for such delay. With respect to delays for which the customer is responsible, however, contracts vary. Some state that regardless of the customer's responsibility, you are not entitled to damages for delay; others either express or imply entitlement to monetary relief.

Assuming that you have a valid basis for a compensable delay claim, you may typically seek to recover costs in the categories that follow.

Wage Escalation. Most construction contracts are labor-intensive and therefore the contract price is based to a great extent upon a calculation of hourly wage rates for labor. If wage rates are increased during a delay through union negotiations, cost of living increases, or other means, you should make a demand for compensation for these extra costs. An example of the manner of computing wage rate differentials follows:

Estimated Hours		
May	June	July
1,000	1,500	700

Actual Hours				
May	June	July	August	September
500	50	1,000	1,000	650

Note that both estimated and actual hours in this example are the same at 3,200, indicating a pure delay case, with no lost efficiency. Assume further that the original rate of $10 per hour for labor jumped to $12 in July due to a new union contract. The contractor is responsible for all costs through the original July completion date.

Estimated Cost		
May	June	July
$10,000	$15,000	$8,400

Total: $33,400

Actual Cost				
May	June	July	August	September
$5,000	$500	$12,000	$12,000	$7,800

Total: $37,300

Thus, you will have a claim for $3,900 for wage escalation.

Material Escalation. Delayed receipt of information from a customer or architect through untimely notices to proceed, delayed issuance of drawings, or delayed approval of catalogue cuts, samples, or shop drawings, may result in your inability to order material as planned during an earlier portion of the contract period. Under these circumstances, you may be forced to purchase material later than anticipated in a period of rising material costs. The differential between the estimated purchase prices contained in the original bid and the actual expense incurred constitutes a compensable claims item.

Extended Storage Costs. If you are required by the contract to store materials or large pieces of electrical equipment until those materials or pieces of equipment are installed, a delay in the time of installation will increase your storage costs by extending the time of storage. You are entitled to recover these excess storage costs as part of a compensable delay claim.

Excess Equipment Costs. The claim also takes into consideration cost of equipment. As a rule, the charge for equipment owned by your company is composed of two elements: depreciation and maintenance. Depreciation expense represents the prorata allocation of the original cost of the equipment over its useful life. The following formula is typically used for determining depreciation:

$$\text{annual depreciation charge} = \frac{\text{purchase price} - \text{salvage value}}{\text{useful life}}$$

Thus, if a piece of equipment, originally purchased for $500, has a useful life of 5 years and a salvage value of $50 at the end of the useful life, the yearly charge using the above formula for depreciation would be ($500 – $50 = $450)/5 = $90. Accordingly, if this equipment is to be used for a contract which is expected to be completed in two years, the charge of $180 should be made for the equipment in the contract. Similarly, the annual depreciation charge will be applied to any extended job duration.

The costs of rented equipment should be claimed on the basis of the rate charged by the lessor. If this rental rate includes servicing, your company will not be permitted a separate maintenance charge in estimating extra costs.

Field Office Overhead. When a construction project is disrupted or delayed by causes beyond your control, you will have a basis for claiming the *excess* job office expenses and *extended* job office expenses. Excess expenses are increased daily expenses resulting directly from changes or disruptions to the work, and extended expenses are increased daily expenses resulting from delayed completion. Computation of excess field office expenses requires a close analysis of those expenses incurred on a per diem or per week basis prior to the event which delayed or disrupted project performance, compared with the expenses incurred thereafter. This claim item may include such items of cost as extra field office personnel, overtime work, additional supplies and other similar costs.

The computation of extended field office expenses is somewhat less complex than the computation of excess field office overhead. The most direct method is to compute the actual field office costs throughout the entire period of the project and divide that amount by the total number of project days. This formula will yield a per diem actual field office overhead rate which can then be multiplied by the number of days the project was delayed.

A variation on this formula is based upon the actual field office expense during the original contract term divided by the original contract period. Thus, the per diem rate is established as that incurred during the *unadjusted* contract term. This amount is then multiplied by the number of days of delay. Neither formula is scientifically perfect, however, since the actual daily field office overhead during the extended period of performance may be less than the overhead during the original contract term. In addition, delayed completion might be a result of changes which do not place a substantial demand on field office expenses during the job completion phase.

The alternative to these formulas, however, is a less desirable "total cost" type approach. In this method, you would subtract your bid estimate for job site overhead from your actual total field office overhead for the job. The difference would constitute the claim for increased job site overhead. Under this total cost approach you must prove the reasonableness of both your estimate and your actual costs, which is often a difficult task.

Extended Home Office Overhead. General overhead expenses are those incurred by your company in the overall management of business. They are not usually incurred for a specific project and thus cannot be charged directly to each project. To remain operational, however, you must recoup all costs incurred in the course of business, including home office expenses.

The methods of computing extended home office overhead are numerous. The contract itself may contain provisions for computation of overhead on the basis of a fixed percentage, most typically 10 or 15 percent, as an amount to be added to any changes or to account for delay costs. Even when a contract contains such a provision, you may not be limited to this fixed percentage amount. For example, it has been successfully argued in many courts that a 10 to 15 percent mark-up applies only to the added costs for the performance of otherwise compensable extras, but does not limit you for claims based upon disruption or delays. It may be, for a particular project, that this fixed percentage figure may produce the best results for your company, but other formulas are available. The most widely accepted formula for computing extended overhead is based on the Board of Contract Appeals case *Eichleay Corp.*, 60–2 BCA ¶2688. Under *Eichleay*, extended overhead is calculated as follows:

$$\frac{\text{Contract billings}}{\text{Total billings for contract period}} \times \begin{array}{l}\text{Total overhead for}\\ \text{contract period}\end{array} =$$

$$\text{Overhead allocable to the contract}$$

$$\frac{\text{Allocable overhead}}{\text{Days of performance}} = \text{Daily contract overhead}$$

$$\text{Daily contract overhead} \times \text{No. days delay} =$$

$$\text{Amount of extended overhead}$$

There are a number of variations on the above formula. An example of the results achieved through application of the *Eichleay* formula is demonstrated below:

> If the contract at issue has a price of $5,000,000, and the contractor, during the period of contract performance, has total sales of $20,000,000, the Home Office Ratio applicable to the project is:
>
> $$\frac{\$\ 5{,}000{,}000}{\$20{,}000{,}000} = 25\%$$
>
> Thus, if the contractor's total home office overhead is $800,000, that portion allocable to the project is:
>
> $$25\% \times \$800{,}000 = \$200{,}000$$
>
> If total project duration has been 200 days, then the per diem overhead is
>
> $$\frac{\$20{,}000}{200} = \$1{,}000 \text{ per day}$$

A fifty-day compensable delay, therefore, will generate a claim for $50,000 in extended home office overhead.

Caution must be exercised in the use of the *Eichleay* formula: it should not be relied on as an infallible means of recouping contract losses. One reason for caution is that the *Eichleay* formula has been treated as applicable only when the contractor has been delayed by a suspension of work as opposed to being delayed by the disruptive effects of changes or extra work orders. The rationale for this ruling is that the *Eichleay* formula is to be used to calculate overhead which is unabsorbed by the contract billings during a period of suspension when there is no work to be done. Changes and extra work actually provide more billable work to absorb overhead expenses and generally incorporate overhead at a fixed rate. In addition, the *Eichleay* formula has been rejected when the contractor failed to lay a foundation for its use by submitting it without any expert proof that overhead was allocated in accordance with standard accounting principles or that no better method of proving unabsorbed overhead losses existed. A number of recent cases have attacked the use of allocation methods as being unrealistic, or have applied more stringent standards of proof in the calculation of home office overhead. For example, at least two cases involving disputes between general and subcontractors have held that the party claiming extended home office overhead expenses must offer evidence to prove that:

a. The delay caused some increase on home office overhead expense incurred by the contractor; and

b. The contractor was unable to obtain other projects to absorb the unallocated overhead costs.

Both of these cases, however, recognized that recovery of unabsorbed home office overhead was a proper element of damages and the decisions in the cases turned upon evidentiary questions. Therefore, it appears safe to say that the use of a formula, without more proof, is insufficient from an evidentiary point of view and that the burden which the contractor must meet to prove such damages is not a settled issue of law. Contractors may best guard against such rulings by allocating as much of their overhead costs as possible to particular projects through the use of cost codes.

Even if you complete performance of the work on schedule or prior to the scheduled completion date, there still may be a valid claim for delay damages if you could have completed the work earlier but for hindrance or interference by the customer. This principle is illustrated by a New York case where the contractor and the State of New York entered into a highway construction contract. Although the contractor actually finished its work eight months before the contract completion date, various unanticipated difficulties were encountered in the course of the construction, especially with respect to the construction of a bridge which was part of the project. Specifically, the contractor alleged defective design, negligent preparation of specifications, inadequate and misleading bid information, unreasonable delay both in preparing and approving new plans and designs, making decisions, prevention of the proper coordination of work, and various other acts and omissions that delayed and impeded the work.

The owner was responsible for many of the above delays. However, the owner argued that because the work was completed and the job was accepted well before the scheduled completion date, there could be no recovery for increased cost due to delay. In effect, the owner was arguing that the only delay which was compensable would be that which caused the project to extend past the original contract completion date.

The court rejected this argument by stating that the contractor had the right to operate free from needless interference by the owner and, therefore, was entitled to compensation since the contractor could have completed the work ahead of schedule and thereby saved substantial sums of money, if it had not been for the delays caused by the owner.

Thus, you can recover compensation for increased costs when the customer interferes with your progress, even if you finished the project on or before schedule. Various boards of contract appeals have also agreed with this principle even under circumstances where the customer did not know of the planned early completion of the project.

Disruption and Efficiency Loss Claims

Disruption and lost efficiency claims are among the most difficult to assess and present. Such costs are usually not accumulated separately in your project cost records, but may nevertheless be substantial. By using the cost coding system described in chapter three, you may document inefficiency losses in a manner which strongly supports an effective claim presentation. The following explanation and tables show how you, as an electrical contractor, may use such cost codes to present an inefficiency claim arising out of a contract to perform electrical work on a multiple-unit housing complex. There are two graphic illustrations of disruption efficiency (or activate and deactivate) which follow the sample documentation, tables, and graphs. The first chart (Figure 7–1) is a sample of activate and deactivate, while the second chart (Figure 7–2) contains disruption and efficiency with a weekly manloading chart overlaid on top. Both charts are visual illustrations of the extensive amount of disruption which occurred throughout the project and would support your documentation in a claim package.

SAMPLE DOCUMENTATION OF INEFFICIENCY CLAIM USING STANDARDIZED COST CODES

Explanation of Cost Codes Used To Show Activation and Deactivation of Labor

Code 100—Conduit—Activate

The activation of Code 100—Conduit can cover either conduit stub-up work or conduit slab work. The difference between the two involves the types of tools used; for the purposes of this presentation we have used the slab work task since it represents better than 80 percent of the Code 100 work.

LAYOUT AND INFO

The first task involved in slab work is the instruction and layout of work by the foreman to the journeymen who will perform the installation. The foreman must pull all reference drawings to supplement the electrical drawings for layout of the work and for instructing the journeymen. Reference drawings include, but are not limited to, architectual, structural, and mechanical drawings. On occasion, certain manufacturers' shop drawings also will be reviewed for detailed size and location of special equipment. In addition, the foreman will relate to the journeymen any special provisions of the specifications which must be observed to make the installation acceptable.

Next, the foreman will advise the journeymen what tools to utilize, and the location of the tools. The journeymen will be advised as to the type, quantity, and location of materials to be used for this specific task.

If the installation requires the crew to work at more than one incremental task at a time, the foreman will designate which individuals will work together on teams in order to make the most efficient use of the labor. Incremental tasks include power, branch, and special systems conduit installations. The foreman must give specific work instructions to each working team when the overall task requires that a number of teams work simultaneously to achieve the desired end result.

PICK UP MATERIALS AND TOOLS FROM WAREHOUSE

The journeymen drive their truck to the job site warehouse. There they inform the warehouseman of their requirements. The warehouseman completes the order, and the journeymen load the materials and tools onto the truck. They must verify the type and quantity of materials and tools to insure they have all items.

The tools will generally include the following:

- Rigid 300 Power Vise
- Rigid 68 Pony Vise
- Drill Motors
- Hickeys
- Chicago Benders: ½ thru 1 inch
- Chicago Benders: 1¼ and 1½ inches
- Hack Saw
- Hammers
- Reamer
- Dies with Ratches: ½ thru 1½ inches
- Tripod Vises
- Rigid Oiler
- Chalk Lines
- 100 Foot Tape

TRAVEL TO WORK SITE

The journeymen are instructed to report to their assigned work location by the foreman after the instruction period. Depending on the distance to be traveled, the men either walk or ride on the truck. They carry with them their hand tools, which consist of the following:

- 1 Knife
- 1 Ball Peen Hammer
- 1 Center Punch
- 1 Hack Saw Frame
- 1 Pencil
- 1 Plumbbob
- 2 Channel Locks
- 1 Keyhole Saw
- 1 Tri-square
- 1 50 Inch Steel Tape
- 1 8 Inch Cutting Pliers
- 1 Long-nose Pliers

1 Small Screwdriver	1 Volt Tester
1 6 Foot Rule	1 10 Inch Brace
1 Diagonal Cutters	1 Level
1 10 Inch & 14 Inch Pipe Wrench	1 Large Screwdriver

SET UP TIME (TOOLS AND MATERIALS)

The tools and materials are unloaded from the truck and transported to the work site. If the work site is located above the main floor, the Foreman attempts to obtain use of a crane or forklift to hoist the required tools and materials onto the proper floors. If hoists are not available, all tools and materials are hand-carried or hand-hoisted to the work area.

The major tools are set up in a designated location and the necessary temporary electrical service is brought in for hookup. Electrical service is required for the Rigid 300 power vise, the Rigid 68 pony vise, and the electric drill motors.

Code 100—Conduit—Deactivate

PICK UP TOOLS AND MATERIALS

Code 100—Conduit Deactivation begins when the foreman instructs the journeymen to remove all tools, equipment and materials from the work area, and to load same onto the truck. It is also necessary to secure the temporary electric service, and generally clean up the area.

Materials are gathered up, and couplings, fittings, and nipples are placed in cartons. Any scrap material is segregated from the usable material and disposed of as necessary. The bulk of the material consists of full lengths of conduit sizes from ¾ to 2 inches. This is gathered into bulk lots and made ready for transfer onto the truck for return to the job site warehouse.

Tools are picked up, placed back in their carrying cases, and made ready for transfer.

SECURE EQUIPMENT INCLUDING RETURN TO SHACK

All equipment and tools are loaded onto the company truck and returned to the job site warehouse by the journeymen. The truck is unloaded by the journeymen, and the materials and tools are placed in designated areas of the job site warehouse. The journeymen return to the shack or work site as assigned by the foreman.

HUMAN NATURE EFFICIENCY FACTOR

When an unscheduled change of work assignment occurs, a portion of production effort is lost due to the journeymen complaining of the work being stopped; they must gather the tools and equipment when they know the job is not finished; they discuss the confusion of the contractor in not completing a task once started; and they doubt the importance of the new work assignment and compare it to the one that is being deactivated. This tendency is inherent in every human, but varies in degree from person to person.

Code 200—Wire—Activate

LAYOUT AND INFO

The first task involved in the wire code work is the layout of work by the foreman and the instruction to the journeymen who will perform the installation. The foreman must pull reference drawings to supplement the electrical drawing prior to instructing the journeymen. Depending upon the task involved, manufacturers' shop drawings may be used for wire insulation color-coding, and schematic wiring diagrams may be used if the system requires interlocking or special control techniques. Specific identification of wires by means of marking tapes is necessary.

The wire code covers both branch and power, and is divided into the following subtasks:

1. Single Conductor Wire
2. Multiconductor Cable
3. Hi-voltage Cable
4. Terminations

In addition to relating the entire scope of the drawings to the journeymen, the foreman will cover any special provisions of the specifications which must be observed to make the installation acceptable.

Next, the foreman will advise the journeymen of the type and quantity of tools to utilize and of the location of the tools. This may be from the job site warehouse or another work location on the project.

The foreman will instruct the journeymen as to the type, size, color and quantity of wire required and where to obtain same. On this particular project, the wire was located in a warehouse separated from the rest of the electrical materials.

When the installation requires the crew to work at more than one incremental task, e.g., branch wiring and power wiring, the foreman will designate which individuals will work in teams to make the most efficient use of the labor. This will necessitate that the foreman give specific instructions to each team separately.

PICK UP MATERIALS AND TOOLS FROM WAREHOUSE

The journeymen, after receiving information from the foreman as to the type and quantity of tools and materials required, will take a truck and drive to the job site warehouse. There they inform the warehouseman of their requirements. The warehouseman completes the order, and the journeymen load the materials and tools onto the truck. They must verify the type and quantity of materials and tools to insure they have all items.

The tools will generally include the following:

Jetline and Tanks | Ladders
Fish Tape | Wire Strippers

Wire Reels | Small Rope or Line

Vacuum Cleaner

TRAVEL TO WORK SITE

The journeymen are instructed to report to their assigned work location by the foreman after the instruction period. Depending on the distance to be traveled, the men walk or ride on the truck. They carry with them their hand tools. Refer to Code 100—Activate for a listing of personal tools.

SETUP TIME (TOOLS AND MATERIALS)

The tools and materials are unloaded from the truck and transported to the work site. If the work site is located above the main floor, the foreman attempts to obtain use of a crane or forklift to hoist the required tools and materials onto the proper floor. If hoists are not available, all tools and materials are hand-carried or hand-hoisted to the work area.

The tools are set up in a designated location and the necessary temporary electrical service is brought in for the hookup of the vacuum cleaner, etc.

Code 200—Wire—Deactivate

PICK UP TOOLS AND MATERIALS

Code 200—Wire Deactivation begins when the foreman instructs the journeymen to remove all tools, equipment, and materials from the working area, and to load same onto the truck. This includes securing the temporary electric service and generally cleaning up the area.

The various wire sizes are either coiled up or placed back onto the reels; any scrap wire ends are segregated from the usable wire and disposed of as necessary. The wire will be contained on reels, in cartons, or in coils and will be located in a central area and ready to load onto the truck.

Tools are picked up, placed back in their carrying cases if applicable, and are ready for transfer.

SECURE EQUIPMENT INCLUDING RETURN TO SHACK

All equipment and tools are loaded onto the truck and returned to the job site warehouse where the tools are unloaded. The wire is transported to the warehouse specifically used for wire and cable. The journeymen then return to the shack or work site as designated by the foreman.

HUMAN NATURE EFFICIENCY FACTOR

Refer to similar description in Code 100.

Code 300—Distribution—Activate

LAYOUT AND INFO

The Distribution Code covers the following:

Distribution Panels

Lighting Panelboards

Power Panelboards

Transformers

Motor Control Centers

Fused Cutouts

Disconnect Switches

The first task involved is the layout of work by the foreman and instruction to the journeymen who will perform the installation. The foreman must pull all reference drawings to supplement the electrical drawing when instructing the journeymen of the work to be performed. The reference drawings include, but are not limited to, architectural, structural, and mechanical drawings. Certain manufacturers' shop drawings will be reviewed for detailed size and location of special equipment. In addition, the foreman will relate to the journeymen any provisions of the specifications which must be observed to make the installation acceptable.

Next, the foreman will tell the journeymen what tools to utilize and the location of tools. The journeymen will be advised as to the location and quantity of materials to be used for this specific task.

When the installation requires the crew to work at more than one incremental task at a time, the foreman will designate which individuals will work as teams to make the most efficient use of the labor. These incremental tasks include, but are not limited to, the installation of cans, interiors, starters and disconnect switches. The foreman gives specific work instructions to each working team when the overall task involves a number of teams to work simultaneously to achieve the desired end result.

PICK UP MATERIALS AND TOOLS FROM WAREHOUSE

The journeymen will take a pickup truck and drive to the job site warehouse. There they inform the warehouseman of their requirements. This is a fairly time-consuming task since each building contains several different types of panels, starters, disconnect switches, etc. There are different physical sizes for cans and interiors, and different electrical sizes for the starters and disconnect switches. It is important that the proper selections be made to avoid confusion and possible rework at a later time. The warehouseman completes the order, and the journeymen load the materials and tools onto the truck. They verify the type and quantity of materials and tools.

The tools will generally include the following:

Stud Gun	K.O. Punch—Hydraulic
Electric Drill	—½ thru 4 Inches

Electric Hammer	Level
Ladder	Hammers

TRAVEL TO WORK SITE

The journeymen are instructed to report to their assigned work location by the foreman after the instructional period. Depending on the distance to be traveled, the men will walk or ride on the truck. They will carry their hand tools. Refer to Code 100—Activate for a listing of personal tools.

SETUP TIME (TOOLS AND MATERIALS)

The tools and materials are unloaded from the truck and transported to the work site. As previously described, it may be necessary for the foreman to locate a crane; otherwise, tools and materials are hand-carried or hand-hoisted to the work area.

The tools are set up in a designated location and the necessary temporary electrical service is brought in for the hookup of the electric drills, hammers, etc.

Code 300—Distribution—Deactivate

PICK UP TOOLS AND MATERIALS

Code 300—Distribution Deactivation begins when the foreman instructs the journeymen to remove all tools, equipment, and materials from the working area, and to load them onto the truck. This includes securing the temporary electric service and generally cleaning up the area.

The various cans, interiors, starters, fused cutouts and miscellaneous materials are gathered up and placed in a central location for later transfer to a truck.

Tools are picked up, placed back in their carrying cases if applicable, and are ready for transfer.

SECURE EQUIPMENT INCLUDING RETURN TO SHACK

All equipment and tools are loaded onto the truck and returned to the job site warehouse by the journeymen. The truck is unloaded by the journeymen, and the materials and tools are placed in the assigned warehouse areas. The journeymen return to the shack or work site as designated by the foreman.

HUMAN NATURE EFFICIENCY FACTOR

Refer to similar description in Code 100.

Code 400—Finish Work—Activate

LAYOUT AND INFO

The first task involved in the Finish Code work is the layout of work by the foreman and the instruction to the journeymen who will perform the installation. The foreman pulls reference drawings to supplement the electrical drawing for use in instructing the journeymen of the work to be performed. Depending upon the task involved, manufacturers' shop drawings may be used for specific light fixture sizes and mounting techniques.

The Finish Code is divided into the following subtasks:

Wiring Devices, Receptables, and Switches

Lighting Fixtures and Accessories

Florescent Fixtures and Accessories

Incandescent Fixtures and Accessories

Lamps

Ballasts

Plates & Covers

Special Fixture Hanger Systems

In addition to explaining the scope of the drawings to the journeymen, the foreman will also cover any special provisions of the specifications which must be observed to make the installation acceptable.

Next, the foreman will advise the journeymen of the type and quantity of tools to utilize and the location of the tools. This may be from the job site warehouse or another work location on the overall project.

The foreman instructs the journeymen of the type and quantity of finish material required, and where to obtain it. On this particular project, the fixtures were located in a warehouse separate from the rest of the electrical materials.

If this installation requires the crew to work at more than one incremental task, e.g., fixture installation wiring devices, receptacles and switches, the foreman will designate which individuals will work in teams to make the most efficient use of labor. The foreman gives specific instruction to each team separately to insure each understands its scope of work.

PICK UP MATERIALS AND TOOLS FROM WAREHOUSE

The journeymen drive a pickup truck to the job site warehouse. There they inform the warehouseman of their requirements. The warehouseman completes the order, and the journeymen load the materials and tools onto the truck. They verify the type and quantity of materials and tools.

The tools will generally include the following:

Small Tap Wrenches

Rolling Scaffolds

Ladders

K.O. Punch—½ Inch

Wire Strippers

Drill Motors

Stud Gun

TRAVEL TO WORK SITE

The journeymen are instructed to report to their assigned work location by the foreman after the instructional period. Depending on the distance to be traveled, the men walk or ride on the truck. They take with them their hand tools. Refer to Code 100—Activate for a listing of personal tools.

SETUP TIME (TOOLS AND MATERIALS)

The tools and materials are unloaded from the truck and transported to the work site. Again, the foreman must arrange for a crane or hoist as required.

The tools are set up in a designated location and the necessary temporary electrical service is brought in for the hookup of the drill motors, etc.

Code 400—Finish Work—Deactivate

PICK UP TOOLS AND MATERIALS

Code 400—Finish Work Deactivation begins when the foreman instructs the journeymen to remove all tools, equipment and materials from the working area, and to load same onto the truck. This includes securing temporary electric service and generally cleaning up the area.

The various materials to be gathered up include the following:

- Wiring Devices
- Receptacles and Switches
- Incandescent Light Fixtures
- Florescent Fixtures
- Lamps and Ballasts
- Plates and Covers
- Hangers and Supports

Many of the items must be replaced in cartons and the lamps must be handled carefully to prevent breakage. All the materials are placed in a central location for later transfer to the truck.

The tools are also picked up, placed back in their carrying cases, if applicable, and are ready for transfer.

SECURE EQUIPMENT INCLUDING RETURN TO SHACK

All equipment and tools are loaded onto the truck and returned to the job site warehouse by the journeymen. The truck is unloaded by the journeymen and the materials and tools placed in the designated areas of the job site warehouse. The journeymen, then, return to the shack or work site so assigned by the foreman.

HUMAN NATURE EFFICIENCY FACTOR

Refer to similar description in Code 100.

Code 500—Special Systems—Activate

LAYOUT AND INFO

The first task involved in the Special Systems Code work is the layout of work by the foreman and instructions to the journeymen who will perform the installation. The foreman pulls certain reference drawings to supplement the electrical drawing for use in instructing the journeymen of the work specifically to be performed. Depending upon the task involved, manufacturers' shop drawings may be used for color-coding terminations, and schematic wiring diagrams may be used when the system requires interlocking or special control techniques. Specific identification of wires is necessary, usually by means of marking tapes.

The Special Systems Code is classified into the following fields:

Telephone System

Heating and Air Conditioning

Fire Alarm and Heat Detection

Sound and Paging

Television

Clocks

Screen Enclosure

Scoreboard

In addition to relating the scope of the drawings to the journeymen, the foreman covers any provisions of the specifications which must be observed to make the installation acceptable.

Next, the foreman will advise the journeymen of the type and quantity of tools and materials needed, and where they are located.

When the installation requires the crew to work at more than one task, e.g., fire alarm and television, the foreman will designate which individuals will work as teams to make the most efficient use of the labor. The foreman gives specific instructions to each team separately to insure each understands the scope of work.

PICK UP MATERIALS AND TOOLS FROM WAREHOUSE

The journeymen take a pickup truck, drive to the job site warehouse, and inform the warehouseman of their requirements. The warehouseman completes the order, and the journeymen load the materials and tools onto the truck. They verify the type and quantity of materials and tools.

The tools will generally include the following:

Stud Gun	K.O. Punch—Hydraulic
Electric Drill	—½ thru 4 Inches
Electric Hammer	Level
Ladder	Hammers

TRAVEL TO WORK SITE

The journeymen are instructed to report to their assigned work locations by the foreman after the instruction period. Depending on the distance to be traveled the men walk or ride on the truck. They take with them their hand tools. Refer to Code 100—Activate for listing of personal tools.

SETUP TIME (TOOLS AND MATERIALS)

The tools and materials are unloaded from the truck and transported to the work site. If the work site is located above the main floor, the foreman must arrange for a crane or forklift to hoist the required tools and materials onto the proper floor. If one is not available, the tools and materials must be hand-carried or hand-hoisted to the specific area.

The tools are set up in a designated location and the necessary temporary electrical service is brought in for the hookup of the electric drill motors and hammers, etc.

Code 500—Special Systems—Deactivate

PICK UP TOOLS AND MATERIALS

Code 500—Special Systems Deactivation begins when the foreman instructs the journeymen to remove all tools, equipment and materials from the working area and load them onto the truck. This includes securing the temporary electric service and generally cleaning up the area.

Depending upon the particular systems which were being installed, the various materials to be gathered up include the following:

Telephone System	Television
Fire Alarm System	Clocks

Heat Detection	Screen Enclosure
Sound and Paging	Scoreboard

Many of the items must be replaced in cartons, and require careful handling to prevent breakage. All the materials are placed in a central location for later transfer to the truck.

The tools are picked up, placed back in their carrying cases, if applicable, and are ready for transfer.

SECURE EQUIPMENT INCLUDING RETURN TO SHACK

All equipment and tools are loaded onto the truck and returned to the job site warehouse by the journeymen. The truck is unloaded by the journeymen, and the materials and tools placed in the designated areas of the job site warehouse. The journeymen return to the shack or work site as assigned by the foreman.

HUMAN NATURE EFFICIENCY FACTOR

Refer to the description of similar item in Code 100.

Table 1
Total time required, in minutes, to perform each of the activating and deactivating functions described in the Sample Documentation.

Activate

Work Item	Code 100 Conduit	Code 200 Wire	Code 300 Distribution	Code 400 Finish Work	Code 500 Special Systems
Layout and Info	30	30	15	15	30
Pick up Materials and Tools from Warehouse	30	30	90	40	30
Travel to Work Site	10	10	10	10	10
Unload and Setup Time (Tools and Materials)	30	20	20	20	15
	100	90	135	85	85

Deactivate

Work Item	Code 100 Conduit	Code 200 Wire	Code 300 Distribution	Code 400 Finish Work	Code 500 Special Systems
Pick up Tools and Materials	15	15	20	30	15
Secure Equipment including Return to Shack	20	30	15	15	15
Human Nature Efficiency Factor	15	15	15	15	15
	50	60	50	60	45

Table 2
Computation of the total hours spent by foremen and journeymen activating and deactivating.

CODE 100—CONDUIT

1	2	3	4	5	6	7	8	9
Bldg. No.	Code 100 Act/Deact (Minutes)	Conv. to Hrs.	Crew Size	Ext. Hrs.	No. of Act/Deact Events	Total Journey-Man Hrs.	Gen. F'man 0.5 Hr. × Column 6	Foreman 1.0 Hr. × Column 6
1	150	2.50	5	12.50	13	162.50	6.50	13.00
2	150	2.50	5	12.50	13	162.50	6.50	13.00
3	150	2.50	4	10.00	11	110.00	5.50	11.00
4	150	2.50	3	7.50	12	90.00	6.00	12.00
5	150	2.50	2	5.00	19	95.00	9.50	19.00
6	150	2.50	2	5.00	9	45.00	4.50	9.00
7	150	2.50	5	12.50	10	125.00	5.00	10.00
8	150	2.50	5	12.50	14	175.00	7.00	14.00
9	150	2.50	2	5.00	19	95.00	9.50	19.00
10	150	2.50	3	7.50	15	112.50	7.50	15.00
11	150	2.50	5	12.50	7	87.50	3.50	7.00
12	150	2.50	6	15.00	10	150.00	5.00	10.00
13	150	2.50	3	7.50	9	67.50	4.50	9.00
14	150	2.50	2	5.00	24	120.00	12.00	24.00
15	150	2.50	2	5.00	8	40.00	4.00	8.00
16	150	2.50	2	5.00	17	85.00	8.50	17.00
17	150	2.50	3	7.50	23	172.50	11.50	23.00
18	150	2.50	2	5.00	7	35.00	3.50	7.00
19	150	2.50	3	7.50	25	187.50	12.50	25.00
20	150	2.50	5	12.50	19	237.50	9.50	19.00
21	150	2.50	3	7.50	14	105.00	7.00	14.00
22	150	2.50	2	5.00	10	50.00	5.00	10.00
23	150	2.50	3	7.50	19	142.50	9.50	19.00
24	150	2.50	2	5.00	13	65.00	6.50	13.00
25	150	2.50	4	10.00	7	70.00	3.50	7.00
						2,787.50	173.50	347.00

Table 2 continued CODE 200—WIRE

1	2	3	4	5	6	7	8	9
Bldg. No.	Code 200 Act/Deact (Minutes)	Conv. to Hrs.	Crew Size	Ext. Hrs.	No. of Act/Deact Events	Total Journey-Man Hrs.	Gen. F'man 0.5 Hr. × Column 6	Foreman 1.0 Hr. × Column 6
1	150	2.50	2	5.00	7	35.00	3.50	7.00
2	150	2.50	2	5.00	14	70.00	7.00	14.00
3	150	2.50	3	7.50	4	30.00	2.00	4.00
4	150	2.50	2	5.00	1	5.00	.50	1.00
5	150	2.50	3	7.50	6	45.00	3.00	6.00
6	150	2.50	2	5.00	2	10.00	1.00	2.00
7	150	2.50	2	5.00	6	30.00	3.00	6.00
8	150	2.50	2	5.00	4	20.00	2.00	4.00
9	150	2.50	3	7.50	5	37.50	2.50	5.00
10	150	2.50	2	5.00	2	10.00	1.00	2.00
11	150	2.50	2	5.00	9	45.00	4.50	9.00
12	150	2.50	2	5.00	11	55.00	5.50	11.00
13	150	2.50	2	5.00	2	10.00	1.00	2.00
14	150	2.50	2	5.00	0	. 0	. 0	. 0
15	150	2.50	2	5.00	1	5.00	.50	1.00
16	150	2.50	2	5.00	6	30.00	3.00	6.00
17	150	2.50	3	7.50	5	37.50	2.50	5.00
18	150	2.50	2	5.00	3	15.00	1.50	3.00
19	150	2.50	3	7.50	6	45.00	3.00	6.00
20	150	2.50	2	5.00	6	30.00	3.00	6.00
21	150	2.50	2	5.00	12	50.00	6.00	12.00
22	150	2.50	3	7.50	7	52.50	3.50	7.00
23	150	2.50	2	5.00	1	5.00	.50	1.00
24	150	2.50	2	5.00	2	10.00	1.00	2.00
25	150	2.50	2	5.00	9	45.00	4.50	9.00
						737.50	65.50	131.00

Table 2 continued

CODE 300—DISTRIBUTION

1	2	3	4	5	6	7	8	9
Bldg. No.	Code 300 Act/Deact (Minutes)	Conv. to Hrs.	Crew Size	Ext. Hrs.	No. of Act/Deact Events	Total Journey-Man Hrs.	Gen. F'man 0.5 Hr. × Column 6	Foreman 1.0 Hr. × Column 6
1	185	3.08	2	6.16	10	61.60	5.00	10.00
2	185	3.08	2	6.16	16	98.56	8.00	16.00
3	185	3.08	2	6.16	19	117.04	9.50	19.00
4	185	3.08	2	6.16	3	18.48	1.50	3.00
5	185	3.08	2	6.16	3	18.48	1.50	3.00
6	185	3.08	2	6.16	1	6.16	.50	1.00
7	185	3.08	3	9.24	9	83.16	4.50	9.00
8	185	3.08	2	6.16	11	67.76	5.50	11.00
9	185	3.08	2	6.16	4	24.64	2.00	4.00
10	185	3.08	2	6.16	4	24.64	2.00	4.00
11	185	3.08	2	6.16	10	61.60	5.00	10.00
12	185	3.08	2	6.16	8	49.28	4.00	8.00
13	185	3.08	2	6.16	2	12.32	1.00	2.00
14	185	3.08	2	6.16	2	12.32	1.00	2.00
15	185	3.08	2	6.16	2	12.32	1.00	2.00
16	185	3.08	2	6.16	3	18.48	1.50	3.00
17	185	3.08	2	6.16	3	18.48	1.50	3.00
18	185	3.08	2	6.16	1	6.16	.50	1.00
19	185	3.08	2	6.16	3	18.48	1.50	3.00
20	185	3.08	2	6.16	2	12.32	1.00	2.00
21	185	3.08	2	6.16	5	30.80	2.50	5.00
22	185	3.08	2	6.16	5	30.80	2.50	5.00
23	185	3.08	2	6.16	5	30.80	2.50	5.00
24	185	3.08	2	6.16	3	18.48	1.50	3.00
25	185	3.08	2	6.16	10	61.60	5.00	10.00
						914.76	72.00	144.00

Table 2 continued

CODE 400—FINISH WORK

1	2	3	4	5	6	7	8	9
Bldg. No.	Code 400 Act/Deact (Minutes)	Conv. to Hrs.	Crew Size	Ext. Hrs.	No. of Act/Deact Events	Total Journey-Man Hrs.	Gen. F'man 0.5 Hr. × Column 6	Foreman 1.0 Hr. × Column 6
1	145	2.42	2	4.84	8	38.72	4.00	8.00
2	145	2.42	2	4.84	4	19.36	2.00	4.00
3	145	2.42	2	4.84	12	58.08	6.00	12.00
4	145	2.42	2	4.84	7	33.88	3.50	7.00
5	145	2.42	2	4.84	6	29.04	3.00	6.00
6	145	2.42	2	4.84	11	53.24	5.50	11.00
7	145	2.42	2	4.84	8	38.72	4.00	8.00
8	145	2.42	2	4.84	8	38.72	4.00	8.00
9	145	2.42	2	4.84	6	29.04	3.00	6.00
10	145	2.42	2	4.84	4	19.36	2.00	4.00
11	145	2.42	5	12.10	6	72.50	3.00	6.00
12	145	2.42	5	12.10	2	24.20	1.00	2.00
13	145	2.42	2	4.84	4	19.36	2.00	4.00
14	145	2.42	3	7.26	5	36.30	2.50	5.00
15	145	2.42	2	4.84	4	19.36	2.00	4.00
16	145	2.42	2	4.84	7	33.88	3.50	7.00
17	145	2.42	2	4.84	11	53.24	5.50	11.00
18	145	2.42	2	4.84	6	29.04	3.00	6.00
19	145	2.42	2	4.84	16	77.44	8.00	16.00
20	145	2.42	3	7.26	12	87.12	6.00	12.00
21	145	2.42	3	7.26	6	43.56	3.00	6.00
22	145	2.42	3	7.26	8	58.08	4.00	8.00
23	145	2.42	2	4.84	4	19.36	2.00	4.00
24	145	2.42	2	4.84	2	9.68	1.00	2.00
25	145	2.42	5	12.10	4	48.40	2.00	4.00
						989.78	85.50	171.00

Table 2 continued

CODE 500—SPECIAL SYSTEMS

1	2	3	4	5	6	7	8	9
Bldg. No.	Code 500 Act/Deact (Minutes)	Conv. to Hrs.	Crew Size	Ext. Hrs.	No. of Act/Deact Events	Total Journey-Man Hrs.	Gen. F'man 0.5 Hr. × Column 6	Foreman 1.0 Hr. × Column 6
1	130	2.17	2	4.33	5	21.65	2.50	5.00
2	130	2.17	2	4.33	3	12.99	1.50	3.00
3	130	2.17	2	4.33	1	4.33	.50	1.00
4	130	2.17	2	4.33	0	0	0	0
5	130	2.17	2	4.33	0	0	0	0
6	130	2.17	2	4.33	0	0	0	0
7	130	2.17	2	4.33	3	12.99	1.50	3.00
8	130	2.17	2	4.33	0	0	0	0
9	130	2.17	2	4.33	0	0	0	0
10	130	2.17	2	4.33	2	8.66	1.00	2.00
11	130	2.17	2	4.33	3	12.99	1.50	3.00
12	130	2.17	2	4.33	1	4.33	.50	1.00
13	130	2.17	2	4.33	1	4.33	.50	1.00
14	130	2.17	2	4.33	0	0	0	0
15	130	2.17	2	4.33	0	0	0	0
16	130	2.17	2	4.33	0	0	0	0
17	130	2.17	2	4.33	4	17.32	2.00	4.00
18	130	2.17	2	4.33	0	0	0	0
19	130	2.17	2	4.33	3	12.99	1.50	3.00
20	130	2.17	2	4.33	2	8.66	1.00	2.00
21	130	2.17	2	4.33	0	0	0	0
22	130	2.17	2	4.33	0	0	0	0
23	130	2.17	2	4.33	2	8.66	1.00	2.00
24	130	2.17	2	4.33	1	4.33	.50	1.00
25	130	2.17	2	4.33	0	0	0	0
						134.23	15.50	31.00

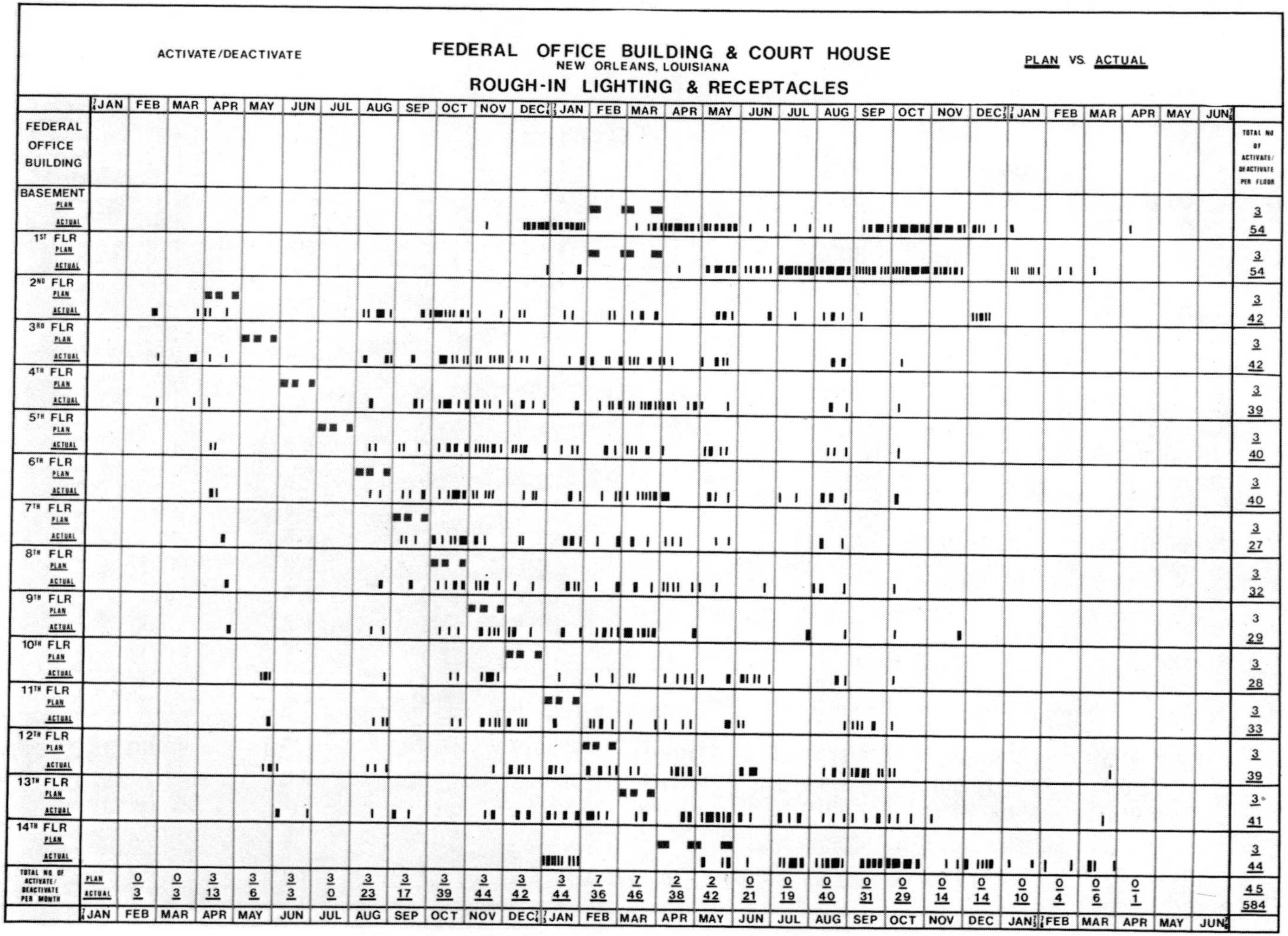

ACTIVATE/DEACTIVATE

FEDERAL OFFICE BUILDING & COURT HOUSE

NEW ORLEANS, LOUISIANA

ROUGH-IN LIGHTING & RECEPTACLES

PLAN VS. ACTUAL

FEDERAL OFFICE BUILDING	JAN	FEB	MAR	APR	MAY	JUN	JUL	AUG	SEP	OCT	NOV	DEC	JAN	FEB	MAR	APR	MAY	JUN	JUL	AUG	SEP	OCT	NOV	DEC	JAN	FEB	MAR	APR	MAY	JUN	TOTAL NO. OF ACTIVATE/DEACTIVATE PER FLOOR
BASEMENT PLAN / ACTUAL																															3 / 54
1ST FLR PLAN / ACTUAL																															3 / 54
2ND FLR PLAN / ACTUAL																															3 / 42
3RD FLR PLAN / ACTUAL																															3 / 42
4TH FLR PLAN / ACTUAL																															3 / 39
5TH FLR PLAN / ACTUAL																															3 / 40
6TH FLR PLAN / ACTUAL																															3 / 40
7TH FLR PLAN / ACTUAL																															3 / 27
8TH FLR PLAN / ACTUAL																															3 / 32
9TH FLR PLAN / ACTUAL																															3 / 29
10TH FLR PLAN / ACTUAL																															3 / 28
11TH FLR PLAN / ACTUAL																															3 / 33
12TH FLR PLAN / ACTUAL																															3 / 39
13TH FLR PLAN / ACTUAL																															3 / 41
14TH FLR PLAN / ACTUAL																															3 / 44
TOTAL NO. OF ACTIVATE/DEACTIVATE PER MONTH	PLAN / ACTUAL	0 / 3	0 / 3	3 / 13	3 / 6	3 / 3	3 / 0	3 / 23	3 / 17	3 / 39	3 / 44	3 / 42	3 / 44	7 / 36	7 / 46	2 / 38	2 / 42	0 / 21	0 / 19	0 / 40	0 / 31	0 / 29	0 / 14	0 / 14	0 / 10	0 / 4	0 / 6	0 / 1			45 / 584

Figure 7–1: Activate/Deactivate Sample

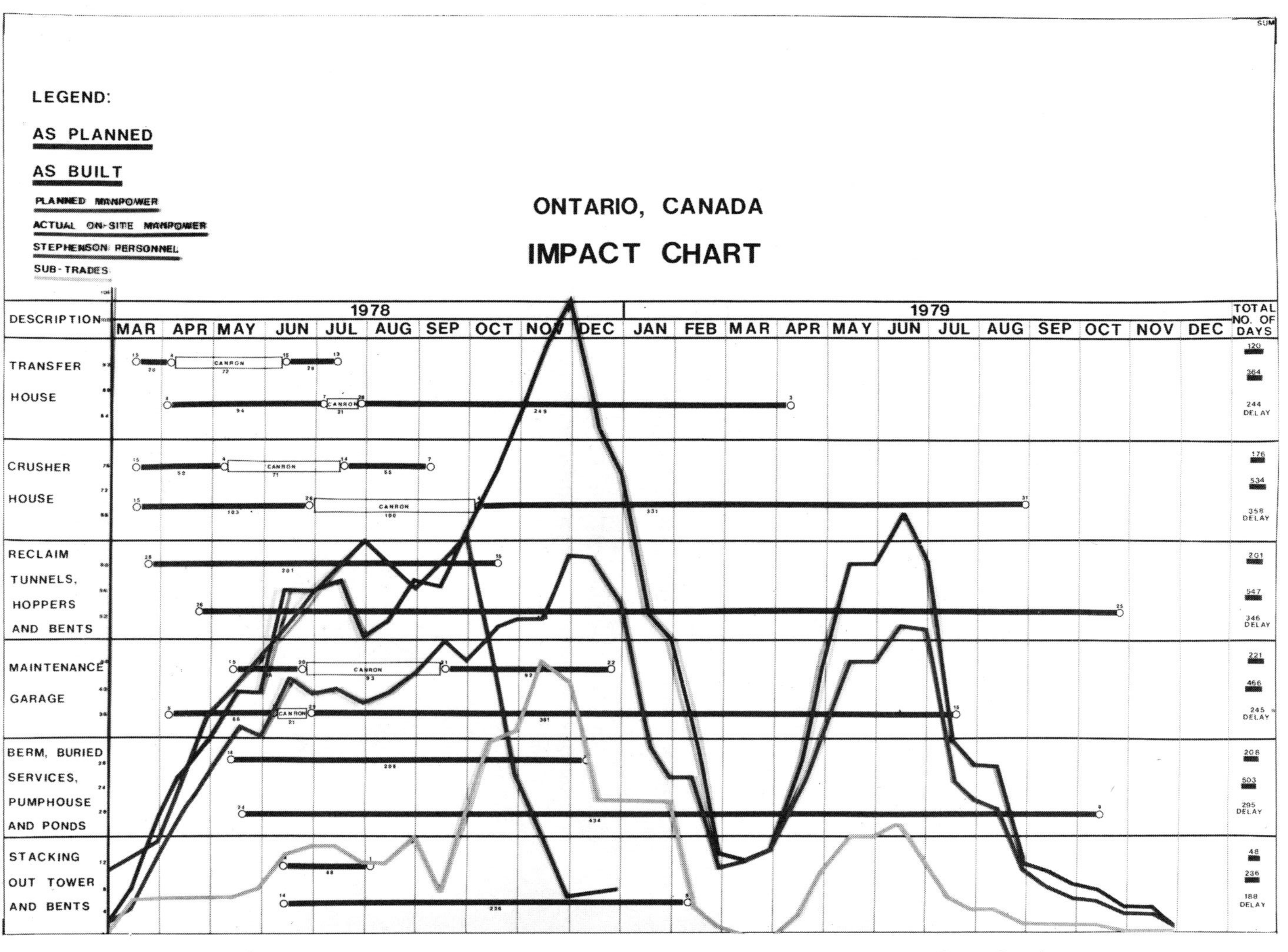

Figure 7–2: Disruption and Efficiency with a Weekly Manloading Chart

If you have maintained careful, detailed records such as those described in the Sample Documentation, it may be possible to segregate actual costs of activating and deactivating labor and other lost efficiency expenses. However, such detailed record-keeping is unusual in the industry, and the costs of maintaining such records may be prohibitive. A total cost approach, which calculates damages as the difference between total actual costs and total estimated costs, is an alternative method of calculating efficiency losses, and is usually accepted by courts and arbitrators in the limited circumstances where no other reasonable method of calculation is possible. A substantial drawback to a total cost formulation is that it places the contractor in the position of having to prove that its original estimate and all of its actual costs were reasonable. Thus, it permits the opposing party to focus upon mistakes, errors, or inefficiencies by the contractor unrelated to the delay or disruption and thus may weaken the overall presentation of your claim.

Expert witnesses' estimates or engineering estimates provide the most acceptable approach for courts or arbitration. Expert testimony on inefficiency costs can be further supported by the total cost method.

Because of the variety of disruption and inefficiency costs, methods of proving them may vary considerably on each project. The following are some of the types of costs which should be taken into account:

a. Equipment inefficiency from downtime;

b. Overtime or premium work required because of disrupted effort or acceleration;

c. Costs related to disruption of planned sequence of operations;

d. Labor inefficiency due to congested work area;

e. Labor inefficiency due to interference of trades;

f. Labor inefficiency due to lack of continuity of operations, improper sequencing of work, working in unfavorable weather, or use of overtime (causing fatigue);

g. Costs related to restricted access to work areas or job site;

h. Costs of increased equipment and manloading (including the hiring of less productive workers to increase manpower and the need to use nonworking or unqualified foremen because of increased crew sizes) required to accelerate performance;

i. Inefficiencies related to unfavorable working conditions such as water on floors of work areas, work performed in less favorable weather conditions than had been anticipated;

j. Inefficiencies related to activation and deactivation of work forces.

Acceleration of Work

Acceleration costs are similar to disruption and inefficiency costs. They may include inefficiencies caused by overtime and stacking of trades, premium wages paid for overtime, additional labor and equipment, additional field supervision, underutilization of supplies and materials, additional field office expenses, additional home office expenses, premiums paid for expediting the delivery of equipment or supplies, and additional transportation expenses.

An acceleration claim may arise under either of two circumstances. In the first case, you may be ordered to complete your work by a date earlier than the contract completion date. This is an express acceleration for which you are entitled to reimbursement by change order for increased costs incurred in complying with this order to accelerate.

In the second case, you: (1) suffer a delay which entitles you to an appropriate time extension; (2) are denied that time extension after making the appropriate request; and (3) then accelerate to meet the original contract completion date. This is a constructive acceleration, and it entitles you to the same measure of recovery as an express acceleration. The law treats this situation as if an appropriate time extension were granted on account of the delay only to be followed by an order to accelerate to meet the original contract completion date.

An acceleration has been recognized to occur if you are required to continue work during a period of abnormal weather conditions during which the project should reasonably have been suspended. Thus, your damages for acceleration included extra weather protection, dewatering operations, premium wages to induce laborers to remain on the job, and other similar costs.

As with disruption claims, expert testimony and engineering estimates may be required to prove damages. Of particular importance in this type of claim is the comparison of planned versus actual manloading of the project based upon the as-built and as-planned CPMs. Manpower curves showing increases in manloading greater than the originally planned number of man-hours can serve as significant graphic evidence of the total effects of an acceleration on your contract performance. The utilization of scheduling analyses to support construction claims is discussed more fully in chapter eight.

Defense of Construction Claims

In general, all the rules regarding preparation and presentation of construction claims are applicable to the defense of construction claims from a converse point of view. Because of the complex series of relationships created among parties to a construction project, each party should be aware of all aspects of both claims presentation and defense.

TECHNICAL LEGAL DEFENSES

Construction contracts are generally complex documents, and they often provide technical legal defenses to construction claims. The number and type of exculpatory clauses which may be used in construction contracts are limited only by the imagination of contract draftsmen and the willingness of contractors to bid on contracts which contain such clauses. Two typical examples of exculpatory clauses which may preclude damages from a party at fault in a large construction claim are notice clauses and "no damages for delay" clauses. As discussed above, you may waive certain claims by failing to notify the party at fault of the existence of a claim within a contractually prescribed period. A "no damages for delay" clause, if found enforceable, may also completely deny you any right to recover damages for delay.

As with notice clauses, however, many courts look with disfavor upon "no damages for delay" clauses because of the harsh result of their application and therefore find ways to avoid enforcing them. Thus, even if the contract contains extensive exculpatory language, the party at fault should not rely solely upon such technical defenses to avoid and defend claims filed against it.

KEEP GOOD RECORDS

The defending party should be just as careful as the claimant to maintain his own daily reports, meeting minutes, job diaries, and correspondence file. Similarly, the defending party should monitor the job for the information which typically indicates that a claim may be arising, in order to avoid being taken by surprise by the claimant.

USE OF CHANGE ORDERS

The change order may be one of the most effective tools available in preventing a construction claim from becoming unmanageable. A change order constitutes a settlement agreement, or an "accord" by the parties which resolves all matters covered within it. Thus, if a party is quick to recognize that it may be liable for a potentially large claim and can persuade the contractor having the claim to execute a change order covering that claim in its entirety, that change order will constitute a valid defense to any further requests for time extensions or monetary compensation arising out of the dispute which it addresses.

Naturally, contractors as well as owners are sensitive to the possible use of change orders to limit recovery for unknown losses. You may deal with this problem by including a reservation of rights to recover costs other than the direct costs of performing the work itself. From the perspective of the recipient of such a proposal, this type of situation is very difficult to handle. A refusal to sign a change order with such reservation will probably disrupt the administration of the project, polarize interests, and lead to a major claim at the end of the work. If the recipient of the proposal does not accept such a reservation, however, it is assuming an open-ended liability for such costs. Therefore, it is recommended that every effort be made to resolve the cost of changes at the time of the change, particularly through the use of the more advanced methods of network planning and scheduling analyses. In the event that no resolution can be obtained, there should at least be some agreement between the parties to the dispute as to what categories of potential impact costs are being reserved.

IMPORTANCE OF SCHEDULING ANALYSIS

If you are faced with a major delay, disruption, or acceleration claim based upon a scheduling analysis, you should have an independent analysis of the claim performed. The claimant's analysis will undoubtedly indicate that actions of the defending party extended the performance of the job by delaying work which was on the critical path. The critical path method of scheduling analysis, however, can be used to defend against, as well as to support, construction claims. An independent analysis could show, for example, that even if the defending party caused certain delays, the claimant may have caused concurrent delays in other work areas which actually caused the critical path to shift and which were the true cause of the overall job delay. A scheduling analysis may also show that because of the claimant's use of float time in the original schedule, the delays upon which the claim is based did not, in fact, cause the job to be completed behind schedule. Finally, a scheduling analysis may well reveal that the claimant's original schedule contained errors in logic or other faults which undermine the basis of the entire claim.

CHAPTER EIGHT

How to Present or Defend a Claim in Negotiation, Arbitration, or Litigation

Introduction

After the claim package is prepared, you should decide which of the alternatives available to you should be taken. At the same time, however, there exists the paramount need and desire for timely economical resolution of these disputes. When a construction contract dispute arises, you have three primary courses of action: you can negotiate a settlement; you can submit the dispute to an arbitrator or panel of arbitrators, providing the opposing party agrees by contract or otherwise; or you can pursue your legal rights and remedies in litigation before a judge or jury. There are relative advantages and disadvantages connected with each of these dispute resolution procedures, and you should examine each in light of your particular circumstances before pursuing a particular course of action. In the evaluation of alternatives, a decision should be based on the maximum amount to be attained by each alternative. Also, a realistic examination of the potential costs of each area should be weighed in this analysis.

Recognizing the trend toward greater claims awareness and construction disputes, this chapter will discuss two important elements of the claim resolution process. First, we will focus on the selection of the dispute resolution procedure which would be most advantageous and effective for your claim presentation or defense. The second section will discuss strategies and techniques for efficient claim presentation or defense. Although there are no fixed rules for either topic, there are certain practical considerations and guidelines which should be kept in mind in order for you to make the most educated and well-informed decision possible under the circumstances of your particular construction dispute.

Try to Negotiate Your Claim

DISADVANTAGES OF NONNEGOTIATED CLAIM RESOLUTION

With greater frequency, parties involved in a construction claim are recognizing and utilizing negotiation as the most efficient and effective method of dispute resolution. This trend toward negotiation of claims has been fostered by three principal forces: high costs, rising inflation and interest.

First, the zeal of an aggrieved party to pursue its claim through the legal arenas of litigation, arbitration, or administrative hearing has been tempered by the increasing high costs of such an action. Similarly, a party involved in a construction claim from either the offensive or defensive standpoint frequently must divert critical and costly management resources to the dispute from the ongoing conduct of its daily business, with the obvious adverse impact on operations.

In addition to the direct costs of legal counsel and the "hidden" costs of lost management time which arbitration and litigation cause, the prosecution or defense of construction claims in any forum has the inherent cost associated with the delay which inevitably accompanies such an action. For example, crowded calendars of federal and state courts, as well as dockets of administrative boards established by federal and state agencies to resolve construction claims, routinely portend a duration of two years or longer between the institution and resolution of a claim—and this estimate does not reflect the potential additional delay possible if a trial court or board decision is appealed by the losing party.

A second principal force contributing to the trend toward negotiation rather than formal litigation of construction claims is rising inflation. With inflation currently in excess of 15 percent annually, claimants envisioning a two-year duration of litigation face a substantial diminution of the value of their claim. This diminution of claim value, however, also serves as an impetus for delay by the party defending the claim with visions of future liability greatly less than the claim amount valued in present dollars.

Construction claimants can also no longer look to recovery of prejudgment interest on a claim as appropriate compensation for delay in claim resolution. For example, even under the limited circumstances where such interest is legally recoverable, the applicable rate of interest in such a situation is the legal rate of interest recognized by federal and state courts of usually no greater than 8 to 10 percent. These rates are clearly no longer competitive with currently available commercial loan or investment rates. Therefore, claimants who must finance a project during litigation have greater attraction to an early negotiated claim settlement.

ADVANTAGES OF NEGOTIATION AS A CLAIMS RESOLUTION METHOD

In addition to the substantial direct and indirect costs and significant potential for delay associated with the formal litigation of construction disputes, there is also a great degree of uncertainty associated with resolution of claims through litigation, arbitration or administrative hearing. For example, a claimant who has unquestionably sustained a substantial monetary loss and thus has a strong equitable claim may find that this claim is denied by a court on a purely legal basis such as lack of timely notice. Therefore, it is beneficial to parties involved in a construction claim to avoid litigation and to pursue a dispute resolution method which offers greater probability of an outcome mutually satisfactory to both parties. For the following reasons, contractors are increasingly utilizing negotiation as a means of successful claim resolution.

Cost Savings

Negotiation ordinarily reduces, and may eliminate, costs incurred by parties for retention of legal counsel or consultants. Although consultants may be required to provide services concerning preparation of a claim package or chart analysis for ultimate presentation at the negotiation, this cost is clearly less than that incurred by a party whose claim is litigated by counsel at a formal proceeding.

Negotiation generally creates less disruption to a company's ongoing operations than litigation. Although a party must devote personnel to a claim both in negotiating and supporting roles, this diversion of corporate resources is generally accomplished in an orderly and efficient manner. A party involved in a court proceeding, however, may be at the mercy of an adversary who subpoenas numerous corporate personnel for depositions and disrupts business with requests for document review and other discovery.

Avoidance of Delay

There is generally a substantial delay associated with litigation in any forum. Negotiation generally increases the potential for dispute resolution in a timely manner. Therefore, you should weigh the value of an early negotiated claim settlement, in a possibly reduced amount, with the value of a successful formal claim litigation, with its resultant costs and delays.

Company Image

A company must also consider its business reputation in all operations. This concern is particularly important under circumstances in which an organization which is continually involved in construction disputes may be viewed as litigious. Negotiation, however, offers a method for you to aggressively protect your interests while allowing dispute resolution on a low visibility level.

Litigation, with its escalation of feelings and costs, may create a long-term adverse reputation for the parties involved. Negotiations, however, if conducted in a businesslike and professional manner, can increase the potential for new or continuing business with the other party.

In the event of an unsuccessful negotiation, it would be a good idea to reevaluate your own position: What was the cause of the failure to settle? Was there unwillingness on *your* part to make reasonable concessions? Is your claim worth the further expense and trouble of arbitration or litigation? Under these circumstances, you will have to decide whether to proceed with arbitration or litigation.

Arbitration or Litigation—The Advantages and Disadvantages

INTRODUCTION

Assuming that diligent good faith efforts to reach a negotiated settlement have failed, you must decide whether to have this dispute resolved by a court of law or by an arbitration panel. This section of the chapter will focus on some important considerations which have an impact on this decision.

Today, contract clauses providing for arbitration of disputes are generally contained in construction industry contracts. If your contract contains such a clause requiring arbitration as a method for dispute resolution, this cause will generally be enforceable if:

a. The dispute arises in a state whose law makes the duty to arbitrate enforceable; or

b. The contract involves interstate commerce and, therefore, the Federal Arbitration Act governs (9 U.S.C. §§1–14); and

c. You have not waived or failed to assert your right to arbitrate.

If there is no arbitration clause or the parties have chosen not to exercise this right, either party can authorize its attorney to initiate an appropriate civil action in a court of competent jurisdiction. For considerations of convenience and tactics, it is generally to your advantage to be the plaintiff and to select the forum. Of course, your selection of a court most likely to favor the merits of your position depends upon the circumstance of your particular case, but any uncertainty concerning whether to initiate the action should generally be resolved in favor of being the plaintiff.

Construction claims arising from multimillion dollar construction projects are frequently litigated in federal court since the nationwide scope of many contractors' businesses and the magnitude of the damages claimed allow federal court jurisdiction upon diversity of citizenship. At the same time, federal court jurisdiction gives the parties the opportunity to utilize the Federal Rules of Civil Procedure, including the new Rules of Evidence, which eliminate the tactical advantages to be gained by a local party and their counsel from familiarity with unusual state court procedural rules. A customer may be more willing to litigate in state court where the waiting period for trial may be shorter and where a local judge or jury may be more sympathetic to the plight of the general contractor. Under either circumstance, you should consider the advantages of litigating the dispute in the vicinity of the project, since access to evidence and witnesses will undoubtedly be an important part of the preparation of your case.

Another important initial decision concerns whether to have this matter decided by a judge or a jury. For example, if your position rests upon a purely technical legal point, such as the customer's failure to deliver customer-furnished material in a timely manner, you may want to submit the case to a judge who may be more sympathetic to your claim of lack of material compliance with the contract than he is to a contractor whose position rests upon a complicated factual scheduling analysis. Similarly, your proof includes evidence of questionable admissibility. In that case you may prefer a judge sitting without a jury, since he is more likely to consider disputed evidence based upon the position that it is better to let in doubtful evidence than to face an appeal.

The nature of the project involved in the dispute is also an important consideration in your choice of a trier of fact. For example, if you are a contractor involved in the construction of a facility such as a school or hospital, a local jury may be very sympathetic to your delay and disruption claims, but may be reluctant to award you damages which will be funded by local property taxes or other special assessments.

There is also a strong feeling in the construction industry that many claims are simply too complex and overwhelming to be comprehended by the average jury. It is equally true, however, that this same criticism may be applied to judges who are equally unfamiliar with construction principles. Under either circumstance, the complexity of the matter and the relative advantage which may be gained by simplification or confusion are important considerations which have a bearing on whether to submit your dispute for decision by a judge or a jury.

TREND TOWARD ARBITRATION OF DISPUTES

Many standard form contracts utilized in the construction industry today include provisions requiring submission of disputes to arbitration. For example, the General Provisions of the AIA Standard Form Contract A-201 (1976) state that disputes will be submitted to arbitration in accordance with the Rules of the Construction Industry Panel of the American Arbitration Association. These rules are included at the end of this chapter for your information. Similarly, disputes arising under international construction contracts are frequently governed by arbitration rules promulgated by the Court of Arbitration of the International Chamber of Commerce.

Unfortunately, too many parties to construction contracts blindly view arbitration as a speedy and efficient panacea for contract disputes. The complexity and magnitude of construction claims, both in dollar and number, which are occurring in construction today, however, demand a more thorough and detailed analysis of the relative advantages and disadvantages of arbitration and litigation as dispute resolution methods.

Speedy Dispute Resolution

The strongest argument historically advanced in support of arbitration is that this dispute resolution procedure is inherently quicker than that available in the crowded judicial system. Arbitration may commence within three months of the filing of an arbitration demand. The trial may not begin until two to five years after the complaint is filed. Even though court proceedings are generally criticized for lengthy delays, it is equally true that arbitration can involve a substantial period of time if one party resists this approach. For example, arbitration can be delayed for a substantial amount of time through any of the following tactics:

a. Formal litigation concerning the scope or enforcement of the arbitration clause;

b. Delay in choosing arbitrators or the location of the arbitration (if not provided in the contract);

c. Unavailability of parties or arbitrators for continuous arbitration sessions;

d. Requests for extensive discovery or prehearing or posthearing procedures; and

e. Failure of the arbitrators to control disruptive and irrelevant cross-examination by counsel.

Therefore, although the intent of arbitration is to allow informal and speedy claim resolution, the nature of the parties and the complexity of issues of a particular claim may reduce or eliminate this supposed advantage of arbitration. For example, it is not unusual for the arbitration hearing of a complex case to extend for more than a year as a result of such factors as the number of witnesses and exhibits, as well as existing commitments of counsel and the arbitrators.

Cost Savings

Similarly, although it is frequently argued that the cost of arbitration is potentially much lower than the cost of litigation, the cost of arbitration depends to a great extent upon the nature of the parties involved and the number and complexity of the issues in dispute. Although parties can arbitrate disputes without being represented by counsel, this circumstance is generally the exception since the complexity of both factual and legal issues and the amount in dispute generally mandate the presence of counsel.

Arbitration can also be as expensive as litigation since counsel will ordinarily utilize the same high degree of preparation for the arbitration of a construction case as for its litigation in court. For example, preparation for arbitration will involve interviews of both lay and expert witnesses and may also involve discovery procedures as later discussed. In arbitrations administered by the American Arbitration Association, the parties are also required to pay compensation to the arbitrators as well as a fee to the AAA depending upon the amount included in the arbitration demand.

Experience and Background of the Arbitration Panel

One of the strongest arguments of the proponents of arbitration is that this dispute resolution procedure allows both parties to have the opportunity to present their positions to a panel of arbitrators who have practical experience and technical knowledge concerning construction, rather than to a jury or a judge with little or no background concerning construction. Since the typical three-man arbitration panel generally includes an architect/engineer, a contractor and an attorney with experience in construction law, it can generally be said that the persons making the arbitration award will be able to draw on their own technical and practical background with the customs and practices of construction. This background is important since arbitrators have the opportunity to ask questions during the hearings which can clarify the complex facts of a dispute by eliciting valuable information. In sharp contrast to this practice, jury members rarely have the ability or opportunity to question witnesses.

Since the AAA maintains an extensive list of arbitrators throughout the country who are qualified in a variety of technical fields, experienced arbitrators are generally available to the parties if they can agree among themselves on the panel. Under the circumstances that they cannot agree on the panel, however, the rules of the AAA provide that the AAA tribunal administrator, who may not be familiar with the particular aspect of construction involved in the dispute, will choose the panel for the parties with resultant loss of potential benefits. For this reason, your review of potential arbitrators should involve considerations not only of the background of the prospective panel member but also the nature of the claim in relation to the background. For example, the arbitration of a controversy arising out of the construction of a nuclear power plant requires experience different from that needed for a dispute arising out of a tunneling contract. If this review is not undertaken by the parties, the arbitration panel may have scarcely more familiarity than a judge or a jury with the constitution procedures and construction law principles which are involved in the particular dispute.

Lack of Legal Technicalities

Arbitrators are not bound by rules of law or evidence, and their powers and scope of injury are limited only by the agreement or order of submission to arbitration, see *Glen Acres, Inc. v. Clifford Corp.*, 228 N.E.2d 835 (Mass. 1967), and *University of Alaska v. Modern Const., Inc.*, 522 P.2d 1141 (Alaska 1974). Since the arbitrators use great discretion in considering what evidence can be allowed in an arbitration hearing, parties are not always limited by rules of evidence or rules of law. Additionally, arbitrators are usually liberal in receiving evidence since refusal to hear evidence, after proper objection, may constitute grounds for vacating or setting aside the award.

The easing or elimination of evidentiary rules, therefore, tends to simplify proceedings and to help avoid long legal arguments over the admissibility of evidence. This approach also allows more expeditious presentation of proof, including documentary evidence and testimony of witnesses. It should be noted, however, that the willingness of arbitrators to receive evidence is not necessarily indicative of the weight or credibility they will attach to a particular item of proof since arbitrators always retain the right to reject or disregard a particular item of evidence presented during the hearing.

In some circumstances, however, the party may wish to rely on the rules of evidence and may find that the arbitrators are unwilling to support such a position. For example, the arbitrators may be willing to listen to hearsay statements even though you might be successful in keeping such statements out of evidence in a court proceeding. It should also be noted that the arbitrators' willingness to hear ordinarily inadmissible evidence can lengthen the proceedings considerably and thus allow a skillful opponent to focus attention on an irrelevant aspect of the dispute, with resultant detriment to your position.

Privacy

One of the strongest advantages of arbitration as a dispute resolution method is privacy. In arbitration, there are no published written decisions nor are hearings open to the public. In this way, there is little danger that the management procedures or financial condition of a party will become a matter of public record through arbitration. This factor can be of great advantage to a party to the proceeding if there are other similar matters pending the outcome of the current arbitration, since there is little chance of an unfavorable arbitration result being used against you.

Lack of Discovery

Usually, there is no discovery in an arbitration proceeding unless specifically authorized by statute, the contract, or other agreement between the parties. Additionally, most state statutes governing arbitration are silent on the issue of discovery. In litigation, a tremendous amount of time and money are required for this kind of pretrial activity, such as exchanges of documents, depositions, and written interrogatories. In arbitration, however, this procedure can often be avoided, resulting in less delay and lower attorneys' fees.

The lack of pretrial discovery generally found in arbitration proceedings may be detrimental to a party. Since pretrial discovery in arbitration is usually within the discretion of the arbitrators, and not a matter of right, the party may not be afforded a proper opportunity to prepare a defense to claims by the other party, see *Commercial Solvents Corp. v. Louisiana Liquid Fertilizer*, 20 F.R.D. 359 (S.D.N.Y. 1957). Several court decisions, however, have shown a greater willingness to permit discovery in arbitration based upon representations by the parties that there exists a special need or unusual circumstance justifying the discovery request, or where the arbitrators have recommended discovery for the parties, see *Vespe Contracting Co. v. Anvan Corp.*, 399 F. Supp. 516 (E.D. Pa. 1975), *Interocean Mercantile Corp. v. Buell*, 207 App. Div. 164, 201 N.Y.S. 753 (1923), and *Cavanaugh v. McDonnell & Co.*, 258 N.E. 2d 561 (Mass. 1970). It should be noted, however, that discovery may be allowed in some jurisdictions even when no special need is shown, provided that it does not delay the proceedings, see *Bigge Crane and Rigging Co. v. Docutel Corp.*, 371 F. Supp. 240 (E.D.N.Y. 1973), applying the United States Arbitration Act. Most courts, however, still require a showing of necessity rather than convenience in those rare instances when discovery is allowed in arbitration.

The arbitrators do retain power to issue subpoenas in arbitration. For example, the United States Arbitration Act contains a provision allowing arbitrators to issue subpoenas to compel the attendance of witnesses and the production of documents, *see* 9 U.S.C. §7. Absent jurisdiction under the federal arbitration act, the issue of the arbitrators' power to subpoena witnesses or documents is a matter of state law, *see* Construction Industry Arbitration Rules (AAA), Section 31.

Unfortunately, adequate discovery in arbitration often results in delays at hearings while parties examine documents or a witness is questioned on each item while counsel "learns" about the document. This procedure may be particularly lengthy if the arbitrators are willing to allow extensive "discovery" questions during the actual hearing. A better approach is to allow a limited continuance for examination of documents, including charts and other demonstrative evidence, that are introduced at the hearing.

Limited Scope of Judicial Review

The losing party in arbitration is quite limited in its ability to appeal an adverse award. The courts will not, as a rule, allow relitigation of a matter which has been previously arbitrated, unless a party can show fraud, bad faith, partiality, or that the arbitrator exceeded his authority, see *McKinney Drilling Co. v. Mach I Limited Partnership*, 32 Md. App. 205, 359 A.2nd 100 (1976). This rule means that even an award based on errors of fact or law will generally be affirmed. In addition, since arbitrators do not have to expose the basis for their decision, no findings of fact are available on which to base an appeal to a court on the merits of the dispute, even if such an appeal route were available. Judicial review of arbitration proceedings is extremely limited, and a careful arbitrator can virtually eliminate any possibility of a successful attack on the results of an arbitration.

ARBITRATION—OTHER IMPORTANT CONSIDERATIONS

Parties

Although construction disputes may involve several parties, they often arise out of the same set of facts or the same occurrence. This unusual feature of construction controversies makes it necessary to consider the feasibility of multiparty arbitration, i.e., consolidation of various arbitration proceedings when all disputes arise out of a common fact pattern. Although consolidation of construction arbitration proceedings seems to offer the advantage of expeditious resolution of disputes arising from a common problem, it is interesting to note that very few construction contracts presently include language making multiparty arbitration mandatory. Rather, it is generally held that mandatory multiparty arbitration cannot be compelled in the absence of a specific agreement between the parties allowing this procedure, AIA Document A–201, Article 7.9.1 (1976 edition); *Stop & Shop Companies, Inc. v. Gilbane Building Co.*, 304 N.E. 2d429 (Mass. 1973). This position is based upon the traditional rule that, without the contractual consent to a multiparty arbitration, a court is without authority to order the consolidation of separate arbitrations, even though they involve the same factual and legal issues, and even though separate arbitrations could lead to inconsistent results. Recent decisions have altered this trend, see *Grover Diamond Associates, Inc. v. American Arbitration Association*, 211 N.W.2d 787 (Minn. 1973) (court ordered consolidation of arbitration after selection of identical arbitrators for owner/contractor and owner/architect disputes); *Vigo Steamship Corp. v. Marship Corp. of Monrovia*, 257 N.E.2d 624 (N.Y. 1970), cert. den. 400 U.S. 819 (1970) (granting New York courts the power to order consolidation of two arbitrations when there are common issues of law or fact); and *Episcopal Housing Corp. v. Federal Ins. Co.*, 255 S.E.2d 451 (S.C. 1979) (owner's claim against general contractor consolidated with the owner's claim against architect when contracts did not preclude consolidation, when there were common issues of fact, and when there would be no resulting prejudice).

It is easy to imagine multiparty situations where arbitration can be a decided disadvantage. For instance, if an owner files a claim against a general contractor for delay damages, the general may desire to join subcontractors in the arbitration, based upon the position that the delay, if any, was caused by the subcontractors. This is not normally possible in arbitration. As a result, the general could find a large award entered against him, and be left to seek out the subcontractors alone by separate long and costly legal proceedings. In recognition of this unjust result, several recent court decisions have construed together the standard AIA arbitration language and subcontract "pass through" clauses to authorize and compel consolidation of arbitration, even when the subcontract provisions contain materially different arbitration clauses from the prime contract, see *Robinson v. Warner*, 370 F. Supp. 828 (R.I. 1974); *Uniroyal, Inc. v. A. Epstein and Sons, Inc.*, 428 F.2d 523 (7th Cir. 1970); and *Chariot Textiles Corp. v. Wanalcin Textile Co.*, 18 N.Y.2d 793, 221 N.E.2d 913 (1966), reversing 21 A.D.2d 762, 250 N.Y. S.2d 493. But, also see *J. Brody and Son, Inc. v. George A. Fuller Co.*, 16 Mich. App. 137, 167 N.W.2d 866 (1969).

One of the countervailing arguments to multiparty arbitration is that such a procedure eliminates the speed and economy of dispute resolution which is one of the basic tenets of arbitration. The increase in the number of parties and counsel will necessarily complicate and have an adverse impact on the expeditious and economical resolution of construction disputes.

Arbitration of International Disputes

During the recent decade, the construction industry has experienced a tremendous growth in the volume of construction abroad. Multimillion dollar building programs, which are being financed by foreign governments as well as foreign private industry, have spurred a tremendous interest in United States-based architect/engineers or contractors to bid and perform contracts overseas.

Dealing with foreign governments and foreign parties may be impeded by unfamiliarity with local legal and economic systems. Contract performance may be affected by political relations between the United States and the foreign government or other governments. These political relations may be an important factor in determining whether a claim by a contractor, architect, or engineer will be favorably treated by a court or administrative tribunal in a foreign jurisdiction. This situation is complicated by the complete freedom of nation states to establish or refuse to establish courts or administrative boards to provide avenues for review of contract disputes. An architect, engineer, or contractor may find himself without relief to enforce payment after services or goods are rendered or supplied. Political turmoil within a foreign country may nullify previous understandings and cause the loss of "golden opportunities" with no remedy available.

The tremendous differences among the legal systems of foreign nations as well as the sometimes justified criticism of the lack of neutrality or impartiality on the part of many foreign courts has prompted a reliance on international arbitration as a process for resolving disputes. Although this process has often been criticized for its tendency to compromise, such an approach has the advantage of providing a framework for the expeditious resolution of business disputes without formality or tremendous cost. An international arbitration is usually conducted by the Court of Arbitration of the International Chamber of Commerce or the Inter-American Commercial Arbitration Commission. Each group has its own rules of procedure which are similar but not identical to the rules of the American Arbitration Association.

Even if a favorable award is eventually rendered, however, a party faces the problem of enforcement of the arbitral award in the foreign country. While treaties often are relied upon for such enforcement, it is a difficult task to achieve such enforcement regardless of the international agreement governing the transaction. Therefore, many United States corporations or governments do not normally seek the assistance of the judicial system but rather rely upon negotiation to resolve construction contract disputes.

CONCLUSION

Thus, arbitration may be the best method of dispute resolution, assuming appropriate contractual language, in cases where a quick decision is desired, where a party does not have a strong legal position, where there are complex technical matters to be decided, and/or where a party is able to present a claim which has the equities on its side. For example, arbitrations involving small monetary amounts, disputes as to contract interpretation (e.g., was the particular item of work part of the contractor's base contract) or sectional disputes (whose responsibility was it to provide a particular item of work—the plumber or the HVAC contractor) are appropriate for resolution through arbitration.

In summary, therefore, you should consider the following factors in comparing the advantages and disadvantages of arbitration and litigation.

1. Experience of the trier of facts
2. Time
3. Discovery
4. Expense
5. Attorneys
6. Procedural rules
7. Rights of appeal
8. Rules of evidence
9. Legal precedents
10. Recovery of damages
11. Multiparty disputes
12. Location of the hearing

In considering arbitration, however, you should be aware that complex claims involving multiple parties and large damages are generally lengthy and costly. Additionally, without a strong, interested, and technically competent arbitration panel, arbitration of complex or multiparty cases may result in a compromise award which is satisfactory to neither party.

How to Persuasively Present or Defend a Construction Claim

INTRODUCTION

The overall goal of any construction claim presentation or defense is to simplify and persuasively present the complex factual and legal circumstances which underly the dispute between the parties.

The achievement of this goal involves the identification of the causes and effects of the performance problems which are the subject of the dispute and the effective communication to the other party of the merits of your position. Since a construction claim ordinarily involves multiple complex factors which are difficult to present, it is easy for you to become confused and diverted in your preparation or defense of a construction claim. This section, therefore, will seek to set forth some tried and true and unique methods for simplifying and effectively presenting or defending a construction claim.

Although you have received considerable information concerning the importance of early claim recognition and documentation, it cannot be overstated that the recognition, preservation and documentation of facts and circumstances of the construction claim are key activities on which you will base your claim negotiation or litigation. Although this book cannot give a foolproof, exact formula for claim analysis and dispute resolution, you must make a concerted, comprehensive effort to gather and organize all documentary evidence (e.g., correspondence, meeting minutes, diaries, photographs, and other project records). You must also carry out your obligations under the contract concerning notice requirements and timely decision-making, in order to put yourself in the best position for early, efficient and economical claim resolution.

As part of the initial "fact finding" phase of claim presentation or defense, you should also consider obtaining an expert opinion concerning the merits of the contractor's claim and possible owner counterclaim. For example, since the owner generally is not expert in construction or design, he will also rely upon the project architect and/or management of the construction project as an initial expert opinion on the merits of the claims and disputes between the owner and the contractor. For example, AIA Document 201, Article 1.5.9 provides that the architect shall interpret the requirements of the contract documents and decide whether the performance of the contractor or the owner complies with these documents. Unfortunately, as a practical matter, the problem often arises under circumstances in which the basis of the contractor's claim is that the performance problem was caused by a design deficiency, while the architect maintains that the problem resulted from defective construction. This commonplace occurrence underlies the importance for the owner to obtain an independent expert who can assist him in deciding which party was primarily responsible for the performance problem. Although many owners and contractors believe that the primary purpose of construction experts is to testify during the litigation or arbitration of construction claims, it is a mistake for you, as the contractor, to fail to employ an expert at an early stage of the claims procedure. The expert's analysis may provide you with the detailed information necessary to avoid lengthy and expensive claim litigation or arbitration.

THE PHILOSOPHY OF SUCCESSFUL CLAIMS NEGOTIATION

For the reasons stated, negotiation usually can be the most efficient and least costly method for resolving construction disputes. The success of this claims resolution procedure, however, depends to a great extent upon your ability to develop a philosophy for negotiating the claim. This section of the materials seeks to convey a philosophy of claims negotiation and includes general guidelines to form the foundation upon which to base the specific tools of negotiation.

Is the Claim Negotiable?

Once a construction dispute arises, the parties may not be capable of mutually negotiating a settlement of the problem. This inability to negotiate may be based on several factors, including both business and nonbusiness reasons. Therefore, a threshold question to the development of a successful negotiation philosophy must be: Is the claim negotiable? To answer this critical question, a party must determine whether an area of mutual interest or concern exists upon which a negotiation philosophy can be founded. There are several analyses which should be pursued in making such a determination.

Determine the Strengths of the Parties' Positions. The initial step of determining the negotiability of a claim is an analysis of the strength of your position as well as your adversary's. This analysis includes a review of several factors, including the following:

1. Merits of the claim. Which party has the stronger argument concerning entitlement to recover damages? The answer to this question will necessarily involve a review of the relevant contract provisions and existing documentation, including project and corporate records.

2. Dollar value of the claim. What are the respective dollar values of the amounts claimed (assuming both a claim and

counterclaim)? This analysis should realistically determine both potential amount of affirmative recovery and possible liability, based again upon a review of the contract and project records. For example, a claim for attorneys' fees may not be recoverable if such a cost item is neither included in the contract nor recoverable under applicable law.

3. Nonbusiness reasons preventing negotiation. There may exist a reason that prevents negotiation of a claim which is unrelated to the actual merits or dollar value of the dispute. For example, personalities of corporate personnel or the desire by a party to delay payment through litigation are two such nonbusiness reasons which make the claim impossible to negotiate.

It cannot be overemphasized that the assessment of the relative strengths of the parties' positions must be a *realistic* and *objective* determination. Occasionally, personal involvement of key personnel of the party or internal corporate politics may lessen the level of objectivity given to this analysis. Possible solutions to this problem include performance of this assessment by corporate personnel not connected with the project which is the subject of the dispute or assessment by outside construction consultants or counsel.

Settlement Authority. The parties may not be able to negotiate a settlement of their dispute because one party lacks authority to settle. Of course, this situation can be overcome if the negotiation is conducted at the correct level of management. Lack of settlement authority may preclude negotiation, however, under circumstances in which one party, such as a public owner, may have to obtain settlement authority from the cognizant local government organization before negotiating the dispute with the general contractor.

Funding for the Settlement. Under current economic conditions, lack of funds has increasingly become an obstacle to claims negotiation. This situation is particularly prevalent in disputes between private parties, although reduction in funding for federal and state construction projects has reduced the amount of public dollars available for funding construction claim settlements. This circumstance places a premium on the creative development of strategies and techniques for arranging a source of funds to effectuate a settlement between the parties.

Compulsory Negotiation. Construction contracts often contain language mandating negotiation of disputes between the parties as a condition precedent to the filing of a claim in litigation or a demand for arbitration. For example, the standard Federal Government Construction Contract requires presentation of the claim to the contracting officer for his final decision and possible appeal to an administrative board. This procedure offers both parties the opportunity for a complete hearing of the merits of the claim and, in practice, settles many disputes. The success of this procedure has also prompted several state and municipal governments to revise their standard construction contract forms to incorporate similar provisions.

Development of the Negotiation Position

Establish Credibility. Credibility is the cornerstone for any negotiation position. This attribute is achieved principally by balancing reasonableness with aggressiveness in presenting and negotiating a claim. Stated differently, you lose credibility if you take a negotiating position which is clearly not supportable by law or fact. For example, a party who claims damages for delay under circumstances where the contract and applicable law preclude recovery of this cost loses credibility.

A party can enhance its credibility through a complete presentation of the relevant facts of the dispute. For example, a claim package presented to the adversary party should necessarily include *all* rather than selective documentation related to a particular issue. Once a party's credibility is reduced, the possibility of a satisfactory settlement substantially diminishes.

Define Objectives. During the course of heated negotiations, it is easy for parties to become temporarily embroiled in collateral issues or personalities. Therefore, it is essential that a party define a set of ultimate objectives as part of the development of a negotiation position. These objectives, which should be confirmed by the management level possessing settlement authority, should necessarily include a "bottom line" settlement cost figure or a narrow range of cost below which the claim cannot be settled without additional approval. Of course, a party should always review and realign objectives during negotiation. Early establishment of objectives, however, ensures that the negotiating team has a coordinated, clear set of ultimate goals.

Develop Strategy. Once the ultimate objectives of the negotiation are defined, you must develop the initial strategy for the negotiation. This effort should encompass decisions concerning what information will be presented, when such data will be presented, and who will be the spokesperson at the negotiation.

There are several factors which a party should consider in developing its initial strategy. These factors include:

1. The adversary negotiation team. Who will represent the adversary? Knowledge of the personalities and expertise of the adversary negotiators is an important tool for strategy development.

2. Level of detail of discussion. Should there be a detailed discussion of facts or summary presentation? Should the presentation stress technical or financial data?

3. Political ramifications. Negotiation, particularly if it involves a governmental or quasi-government party, may have potential ramifications extrinsic to the actual merits of the controversy.

4. Proposal. Should a proposal be put on the table for consideration? It is ordinarily recommended that a party resist making the first settlement proposal which may embarrass or antagonize the opposing party. When a party is seeking substantial additional benefits, however, it may be necessary to set forth a proposed solution for the claim to initiate serious discussion.

In developing this strategy, however, the successful negotiator must remain flexible to alternative approaches to dispute resolution. For example, new facts, policy, or political developments introduced into the negotiation may compel a revision of strategy.

Having developed a negotiation position with appropriate objectives and alternative strategies, a party can approach the bargaining table well prepared.

Negotiation Tools and Techniques

Be Prepared. Preparation is an essential prerequisite for successful negotiation. This task involves a comprehensive marshalling and mastering by the negotiating team of all facts and figures relevant to the negotiation. There is nothing more embarrassing and detrimental to a party's credibility and overall negotiation position than lack of familiarity with the factual and legal background of the claim.

Many parties have utilized the preparation of a claim book as a method of organizing and presenting its position on the disputed issues. This document, which ordinarily sets forth the factual, financial, and legal bases of a party's claim, can either be presented to the other party or used as a reference document. The claim package can also be used as a briefing document if litigation between the parties subsequently ensues.

As an alternative or complement to the claim document, a party may prepare appropriate charts or graphs to better illustrate its claim position. This approach is particularly useful under circumstances in which the claim involves delays or disruptions to the project schedule.

Negotiate as a Team. The complex and extended nature of most construction claim negotiations makes it inadvisable, if not impossible, for a single negotiator to successfully represent a party's interest. Therefore, it is wise to establish a team of several individuals to negotiate the construction claim.

Although the actual negotiating is generally done only by a team spokesperson, the team approach has several advantages. For example, the lead negotiator can rely on team members to "critique" the negotiation and to offer helpful recommendations for revision of strategy. The team approach also creates the potential for a "white hat–black hat" negotiating technique. This technique ordinarily either involves alternate role playing by different members of the team, or it creates the possibility for a party to introduce a third person (such as an upper level manager) into the negotiation as a "fall guy." Finally, the team concept allows a party to utilize both project and home office personnel to present its position. This flexibility of personnel may be particularly advantageous under circumstances where adverse feelings may exist between respective project personnel as a result of the pending dispute.

Establish the Correct Entry Level for Discussion. A delicate issue faced by any negotiating team is assessment of the correct level of the adversary party's organization on which to commence formal claims negotiation. Frequently, this issue can be answered by a determination of the management level which has authority to settle the dispute. Under circumstances where a dispute involves a governmental or quasi-governmental party, however, political conditions may dictate initiation of negotiation at a management level different from that of the manager with settlement authority. A party can generally determine the correct entry level for claims negotiation through informal contact between respective lower level project management personnel.

Forum. A party who possesses the "home field" has a psychological as well as tactical advantage in negotiation. The ability to negotiate in familiar surroundings, the proximity of support personnel, and convenience are assets which accrue to the party negotiating on its own turf. At the least, a party should insist on a neutral site for all formal claim negotiations.

Format of the Discussion. An important attribute for a successful negotiation is the ability to direct the discussion to avoid diversion to collateral issues and to lessen conflict of personalities. This goal can best be achieved through development and use of an agenda and timetable for each negotiation session. Such a procedure also has the advantages of creating the opportunity for consultation among the negotiation team and with higher management personnel and of preventing marathon negotiation sessions which often become counterproductive.

The importance of a social respite from the bargaining table for both parties also should not be overlooked. Such events as meals and cocktail parties scheduled during an extended negotiation give both parties the opportunity to establish a better rapport with each other and may have a resultant positive impact on negotiations.

Assist the Adversary. You may often discover that the key to resolution of the claim is to assist your adversary in resolving his own "in-house" problems, or overcoming problems to settlement created by third parties. For example, in asserting a valid claim for delays against the general contractor, you might suggest to the customer that delays by other subcontractors may create liability for them for a share of the damages assessment. As another example, a party may only be able to effectuate a claim settlement after resolving issues outstanding with a government regulatory agency which has an interest in the dispute. Both parties may be jointly required to provide information to government agencies to obtain licenses or permits which may be necessary for the settlement agreement.

In presenting a draft document for resolving the claim, you should attempt to frame the proposal in the light most favorable to your adversary. This task may require a party to be flexible in the format of the resolution document. For example, political or policy considerations may require that the settlement terms be reflected in a Memorandum of Understanding while the financial information be included in a separate contract document.

In drafting such a resolution document, it is also important to remember that each party should be able to leave the bargaining table having gained benefits. Stated differently, the prevailing party in the negotiation should attempt to confer some benefit, whether monetary or not, on its adversary as part of the agreement. This benefit frequently takes the form of agreements to perform future additional services or furnish additional equipment, promises for new business relationships, or other mutual promises which may make the settlement more "face saving" for the "losing" party.

FORMAL CLAIM PREPARATION OR DEFENSE

If the parties are unable to negotiate a settlement of the dispute, you and your counsel or consultants must begin to prepare for formal claim presentation or defense. Although the forum for the dispute resolution may present different strategy considerations and procedural rules, there are several common essential elements for formal presentation or defense of a claim to a judge, jury, or arbitration panel. This task involves the collection, organization, and evaluation of the mass of facts, records, and opinions that form the basis of a construction claim into a simple and coherent presentation which supports your legal and equitable claim position, establishes liability, and proves damages.

Although thorough preparation and effective presentation techniques are similar for the presentation or defense of construction claims in arbitration or litigation, there is one essential difference. For the reasons stated above, arbitrations generally do not involve pretrial procedures such as discovery and prehearing conferences, although these procedures are available in formal litigation.

Discovery procedures, such as depositions, interrogatories, requests for admission, and pretrial conferences, have the advantage of offering both parties the opportunity to narrow the focus of inquiry by eliminating extraneous issues or obtaining agreements such as stipulations of fact or authenticity of documents. For example, the pretrial conference offers the parties the opportunity to review with the judge unusual legal issues involved in the particular case which can help both parties to ensure the most efficient presentation of proof at the actual hearing. For example, a determination concerning the legal measure of damages recognizable by the court, if it is the subject of a prehearing conference, can help you make a close analysis of the planned method of proving such losses as delay damages. Similarly, the prehearing conference can resolve potential legal issues such as lack of written notice or applicability of exculpatory clauses (e.g., no damages for delay, hold harmless) before trial, with resultant reduction of trial days and legal fees.

As part of arbitrations administered by the AAA, it is also possible to request a prehearing meeting with the arbitrators for purposes of legal rulings. This procedure is infrequently used, however, and may be frustrating to counsel dealing with indecisive arbitrators who are unwilling to decide these matters prior to the actual hearing.

Of course, the most difficult task confronting you and your counsel is the presentation of evidence to the arbitration panel, judge, or jury as a means to present or defend the construction claim. Although the choice of tactics and strategies in either forum is the prerogative of the party and its counsel based upon the strengths and weaknesses of the position, it is generally best to present the two broad issues of liability and damages separately. At the same time, however, the ability of the magnitude of damages suffered by the party to influence the issue of liability may prompt a party and its counsel to resist bifurcation of these two general issues.

Liability

Contract Rights and Obligations. The issue of liability in construction contract claims principally involves a comparison of the rights and obligations of the parties to their disputed actions during the course of contract performance. This analysis generally entails a review of the express contract responsibilities as well as implied obligations of each party (e.g., duty not to hinder, duty to cooperate) and the set of facts which allegedly may constitute a breach of these obligations.

As a practical approach to reviewing a party's position on the issue of liability, the construction attorney should compile a checklist of the rights and obligations of the client annotated with relevant references to factual circumstances, including appropriate documentation, which bear on this issue. This checklist will provide an outline for a party's order of proof, including documentary evidence, on this issue.

Witnesses. The presentation of your position on liability must be principally accomplished through live testimony of witnesses. During trial preparation, therefore, the construction attorney must interview all prospective witnesses to determine the depth of their first-hand knowledge of the facts as well as to apprise each potential witness of the overall position of the client and of the witnesses' particular role in the claim presentation or defense.

Any counsel or party who has participated in the trial or arbitration of a major construction case recognizes that the most important client personnel in the presentation or defense of a construction claim are the individuals, usually at middle or low management level, who have been involved on a day-to-day basis with the construction project. Although the executives of the owner or architect may have general knowledge concerning overall performance problems or disputes, it is ordinarily the owner's on-site representative, project architect or other "field level" individual who will provide the detailed, in-depth testimony upon which any successful claim presentation or defense is based. These individuals, who frequently have colorful personalities, also offer the potential for keeping the attention of the judge, jury, or arbitration panel by simplifying sophisticated construction problems into ordinary and understandable descriptive testimony.

By its nature, a construction case involves the presentation of numerous facts. Therefore, the logical sequence of witness testimony is a chronological one. For example, the first witness should be that individual who was involved with the preconstruction and award phases (whether bid, negotiated, or otherwise awarded). Later witnesses follow the chronological order of proof, with testimony concerning the progress of construction and particular performance problems (shop drawing submittals, changes, differing site conditions, etc.).

In developing a list of witnesses for trial, the construction attorney must also consider the necessity of hiring experts to participate in the proof of the case. Traditionally, construction claim consultants have performed several functions for trial counsel including: assemble, review, summarize and analyze project records and documentation; appraise quality of work; assist trial counsel in understanding the technical elements of the dispute, including customs and practices in the particular trade; prepare schedule and cost analyses to analyze impacts of interferences to the project schedule; and be available as expert witnesses.

It is important to give high priority to engaging a claims consultant *early* in the construction dispute. The timing of this decision is particularly important since the claims consultant can be of great assistance in identifying the strengths and weaknesses of the position of both your client and its adversary. The consultant should obtain as much first-hand knowledge of the project as possible in order to avoid having to base an expert opinion solely upon documentation and interviews with project personnel. A claim consultant can also assist in the documentation of the factual aspects of the disputes, including establishment of proper record-keeping procedures for such important areas of the project as schedule and cost control.

Since construction claim consulting is a burgeoning profession, it is important for your counsel to hire experts who have the best available credentials for the specific needs of the particular dispute. The criteria for selection of experts is to a great extent similar to criteria discussed earlier in this article concerning selection of arbitrators. The construction attorney should look at the education, construction experience, scope of knowledge, and previous consulting claim experience in the context of the particular facts of your case. For example, counsel for a contractor of a metropolitan sewerage authority should look for consultants with previous experience with claims arising out of wastewater treatment projects, including phased scheduling of contractors, EPA-grant funding requirements, and experience as expert witnesses in particular technical areas such as cost or damages analysis. This review should include contact with references offered by the claim consultant as well as an introductory meeting with those personnel from the claim consulting firm who will work on your particular case. In choosing a construction claim expert, the construction attorney should also be aware that he is the quarterback of the trial team and he should be reluctant to engage a consultant who may attempt to "take over the case."

Documentary Evidence. Particularly in construction litigation, use of documentary evidence is vital to prosecuting or defending a claim. Of course, the kind and number of items offered into evidence depends upon the quantum of proof required to defeat or prove a claim. Your review of the pleadings of the parties will usually indicate this quantum. For example, the proof of a single fact, such as the date of an event, may be established with one document. If a party is required to prove that it gave the notice required by the contract, however, it may have to introduce ten documents such as correspondence, meeting minutes or other documents to make a prima facie case.

To a certain extent, an attorney is compelled to introduce large numbers of documents in construction cases, both to set forth the disputed issues and to build a record. Unfortunately, there is real danger in engaging in a "paper war" in construction litigation. First of all, this tactic frequently obscures the critical points that a party wants to set forth. The trier of fact may also become confused if a mass of documents is offered into evidence. These considerations prompt great discretion and selectivity in determining the number and sequence of documents to be presented so that evidential overkill will not destroy an otherwise meritorious position.

Use of Summaries and Charts. When statistical information, such as labor costs or daily manpower reports, is required for the proof of your case, it is wasteful and confusing to attempt to present this evidence through the introduction of hundreds of documents. Therefore, you should consider presenting this voluminous information in an attractive summary chart or graphic form which, coupled with testimony, is a persuasive method of proof.

The Rules of Evidence permit the use of charts, summaries, or other demonstrative evidence at trial, see Federal Rule of Evidence 1006. The prerequisites for the use of this evidence are: (1) that the chart or summary contains only information drawn from original or duplicate documents, *Pritchard v. Liggett Myers Tobacco Co.*, 295 F.2d 292 (3rd Cir. 1961), and that the party seeking to introduce the chart must make these original or duplicate records available to the adversary for review and copying, *Nichols v. Upjohn Co.*, 610 F.2d 293 (5th Cir. 1980). Of course, the trial court has discretion concerning the use and admission of such charts and other demonstrative exhibits, *United States v. Brickey*, 426 F.2d 680 (8th Cir. 1970).

As a practical matter, the construction attorney seeking to introduce a summary or chart exhibit should also offer into evidence the original or duplicate documentation since the summary itself is not admissible [and therefore cannot be considered by the jury in deliberations], unless the documents which are summarized are in evidence or available at the trial, see *Melinder v. United States*, 280 F. Supp. 451 (W.D. Okl. 1968).

In presenting documents, it is also an excellent trial tactic to utilize your opponent's records to the greatest extent possible. This is particularly appropriate and possible in litigation where discovery procedures will allow a detailed and complete review of your adversary's project records. For example, to the extent possible, any summary of labor reports, project costs, or manpower should be based upon your opponent's records rather than your own in order to increase the credibility and overall strength of your proof.

Delay and Disruption Claims

The Contractor's Burden of Proof. In documenting claims, it should be remembered that you have the burden of proof to establish and isolate the period of delay attributable to the customer or its agents. It is only after this liability for delay or disruption has been proven that you are allowed to prove the damages which it incurred during the period of delay or disruption. It is important to note that it is an often cited principle of construction law that courts will not make any effort to apportion damages in a situation where both parties are found to have contributed to the delays in contract completion, *United States v. United Engineering Contracting Co.*, 23 U.S. 236 (1914); *Commerce International Co. v. United States*, 167 Ct. Cl. 529 (1964).

As part of your burden of proof, therefore, you will be required to prove that the customer was responsible for "delay," i.e., an act or occurrence which prevents the actual performance of an activity or group of interrelated activities (e.g., completion of a project) from taking place as scheduled. It is important to note, however, that delay in a single activity does *not* necessarily mean delay in completion of the project, despite common misconception see *Dawson Construction Co., Inc.* 75–2 BCA ¶11–563 (1975); *Sovereign Construction Co., Ltd.*, 75–1 BCA ¶11–251 (1975).

In support of the claim, you will also be required to prove that the delay to the work was excusable (e.g., resulting from changes, differing site conditions, lack of site access, unusually severe weather or strikes) rather than nonexcusable (e.g., resulting from your failure to properly schedule the work or furnish sufficient work or materials, lack of coordination, or lack of timely shop drawing submittal) in order to prevail on a delay claim.

The Role of Schedule Analysis. You should generally base your determination of the delay upon a comparison of your actual performance (whether of a single activity or completion of a project) and a particular performance standard. This analysis will ordinarily focus upon a comparison of the as-planned and as-built conditions of the project. An example of this type of analysis is included as Figure 8–1 herein. The performance standard or as-planned schedule is generally the original project schedule. This document, which is

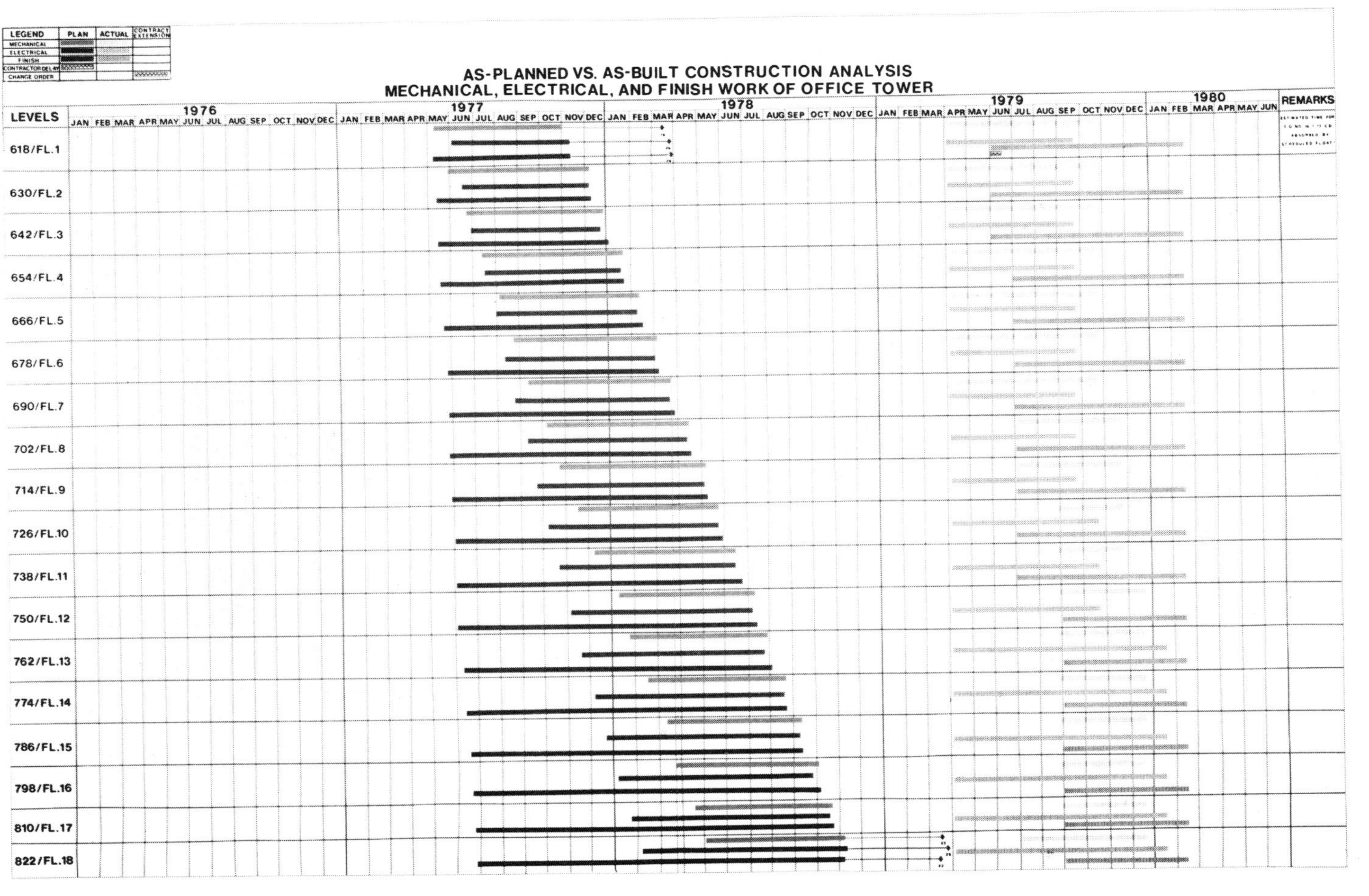

Figure 8–1: As Planned and As Built

usually prepared by you, can be in one of several forms, e.g., bar graph, critical path method schedule, or other network scheduling method. This schedule represents how the project should have progressed had no problems been encountered.

There is generally no dispute concerning your as-planned schedule since it is ordinarily approved by the owner or the architect/engineer. If this is not the case, however, the owner and its counsel will attempt to discredit your as-planned schedule utilizing any or all of the following arguments:

a. The schedule does not reflect a reasonable sequence of work;

b. The schedule is not based upon sound logic;

c. This schedule is not mathematically correct;

d. The schedule does not conform to the project records; or

e. The schedule does not account for all relevant delays.

See *Chaney and James Construction Co. v. United States*, 190 Ct. Cl. 699 (1970); *A. Teichert and Sons, Inc.*, 68–2 BCA ¶7175 (1968); *Joseph E. Bennett Co.*, 72–1 BCA ¶9634 (1972).

The as-built schedule is a model of the project as it was actually performed. This schedule can be prepared during construction or after the fact. Nevertheless, if based on reliable data, the as-built schedule is a true depiction of the project performance. In analyzing your as-built schedule, it is important to review the documents used in establishing dates of performance. Since it is not always easy to actually reconstruct the progress of work on a particular project, contractors and owners have traditionally used some or all of the following information to perform this analysis:

a. Daily reports of the owner or contractor;

b. Requisitions for progress payments on a monthly or other periodic basis;

c. Progress meeting minutes;

d. As-built contract drawings indicating the dates of performance of various portions of the work, such as concrete pours;

e. Correspondence;

f. Progress photographs; and

g. Change orders.

The owner and its counsel, therefore, will independently verify the basis of your as-built information from available project records.

The critical focus of your delay claim is the so-called as-adjusted schedule which is a model of how the original project schedule was affected by excusable delays including both compensable and non-compensable delays. This schedule analysis technique is of considerable use in claims presentation and has been recognized by various courts and boards of contract appeals, see *Blackhawk Heating and Plumbing Co.*, 75–1 BCA ¶11,261 (1975). The goal of this analysis is to introduce into the as-planned schedule the delays, disruptions and other schedule interferences which form the basis of your claim. By doing this, a new schedule is constructed which represents the manner in which the project as originally planned should have progressed, taking into account all the performance problems. The result is a model showing the amount of additional time, if any, you should have been allowed in the construction of the project.

The success of this as-planned–as-built–as-adjusted schedule analysis depends to a great extent upon the kind of schedule utilized for this analysis. For example, if the contractor utilizes a bar graph for this scheduling analysis, it can easily be attacked on the grounds that it does not and cannot show interrelationships between activities as was discussed in chapter five. Secondly, a bar chart does not indicate which activities are critical to the timely completion of the project. On most construction projects, a relatively small number of activities are critical to the completion of the project and the remaining activities can begin and finish on any one of several days without impacting the timely completion of the project. Thus, it is vitally important to monitor the activities which cannot be accomplished utilizing a bar chart.

Courts and boards have recognized the limited value of bar charts in claim presentation in such cases as *Minimar Builders, Inc.*, 72–2 BCA ¶9599 (1972). In that case the Armed Services Board of Contract Appeals recognized the inherent limitations of a bar chart when it stated:

> Although two of appellant's construction schedules were introduced in evidence, one which had been approved by the Government and one which had not, neither was anything more than a bar chart showing the duration and projected calendar dates for the performance of the contractual tasks. Since no interrelationship was shown between the tasks the chart cannot show what project activities were dependent on the prior performance of the plaster and ceiling work, much less whether overall project completion was thereby affected. In short, the schedules were not prepared by the Critical Path Method (CPM) and hence are not probative as to whether any particular activity or group of activities was on the critical path or constituted the pacing element for the project.

Therefore, in recognition of the limited usefulness of bar charts in claims preparation and presentation, contractors and owners are more frequently utilizing network scheduling techniques, such as the critical path method, as the basis for their claims. The use of the network techniques as evidentiary tools has been recognized by both courts and boards, see *Brooks Towers Corp. v. Hunkan-Conkey Construction Co.*, 454 F.2d 1203 (10th Cir. 1972); *Raymond Contractors of Africa, Ltd. v. U.S.*, 198 Ct. Cl. 147, 411 F.2d 1227 (1969).

Your problem, therefore, is to identify the critical path through the job and relate the owner and prime activity to the path. It must be shown that the delays or disruptions complained of affected the critical path, and any delays or interruptions to the work as a result of actions of you, as the contractor, or other causes which are not the owner's responsibility affected only noncritical items. (See Figure 8–2 for a graphic illustration of as-planned and as-built with delay and disruption areas shown as they affected job progress.) Stated differently, you must successfully answer the question: But for the actions of the owner or its agents, when would the project have been completed? If you fail to segregate nonrecoverable causes of delay or disruption, you risk the denial of your claim, see *Lichter v. Mellon-Steuart Co.*, 193 F. Supp. 216 (W.D. Pa. 1961); *aff'd.*, 305 F.2d 216 (3rd Cir. 1962).

When the customer is analyzing your delay or disruption claim based upon a CPM analysis, the customer will focus on four general defenses. The most obvious defense is that no delay to the project occurred. For example, although project correspondence may indi-

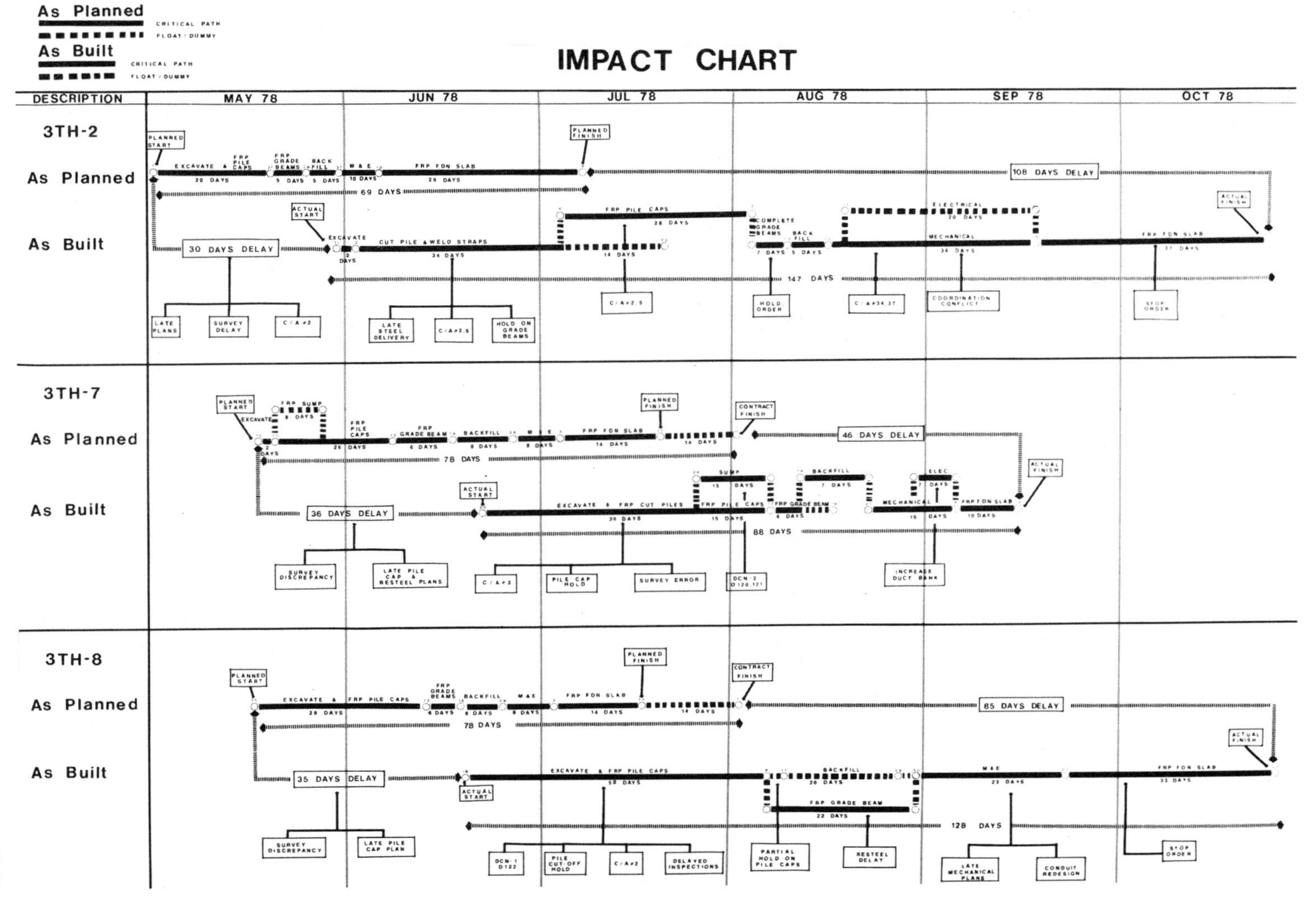

Figure 8–2: As Planned and As Built with Delay and Disruption Areas

cate notice of an impending delay, review of project records including your as-built schedule may indicate that no delay occurred.

The use of a critical path method analysis also allows the owner, contractor, and their counsel to determine whether in fact the delayed activity was on the critical path or not. For example, Figure 8–3 and Figure 8–4 indicate an analysis of the effect to the project critical path of changes ordered by the owner during the construction of an office building and cultural center. This chart focuses on the critical path of the project which went through the office tower of the project with secondary paths (including float) located in the music theatre and other areas of the cultural center. This analysis was utilized to determine whether the delayed activity was on the critical path or became critical. It should be noted, however, that if a delayed activity does become critical, the extent of the delay in completion date is not the same as the delay to the activity. For example, Figure 8–3 indicates that a delay in the drama theatre could be as long as 478 days without impacting the overall completion date of the project.

The most widely used defense based upon the critical path analysis is concurrent delay. In raising this defense, the owner will try to concentrate on introducing evidence that a delay to project completion resulting from an owner-caused condition, such as changes in the work, occurred at the same time the project was being delayed because of a condition which was your responsibility (e.g., lack of equipment or manpower). To support this position concerning concurrent delays, the owner will review as-built project documentation, including your records. This latter source of information is particularly important since many general contractors will take inconsistent positions, vis-a-vis the owner and a subcontractor. For example, the general contractor may claim that a delay results from defective plans and specifications furnished by the owner, while at the same time he may claim that the item of work which was the subject of the claim was installed by a subcontractor in a defective and unworkmanlike manner.

You, as the contractor, therefore, should take full advantage of the critical path method progress schedule or other network analysis to support your claim. CPM analysis does provide a dynamic evidentiary tool which is especially useful in graphically displaying cause and effect. It should be remembered, however, that successful support of your claim results from both proper presentation of your position as well as understanding of how the owner will use the CPM to analyze your claim.

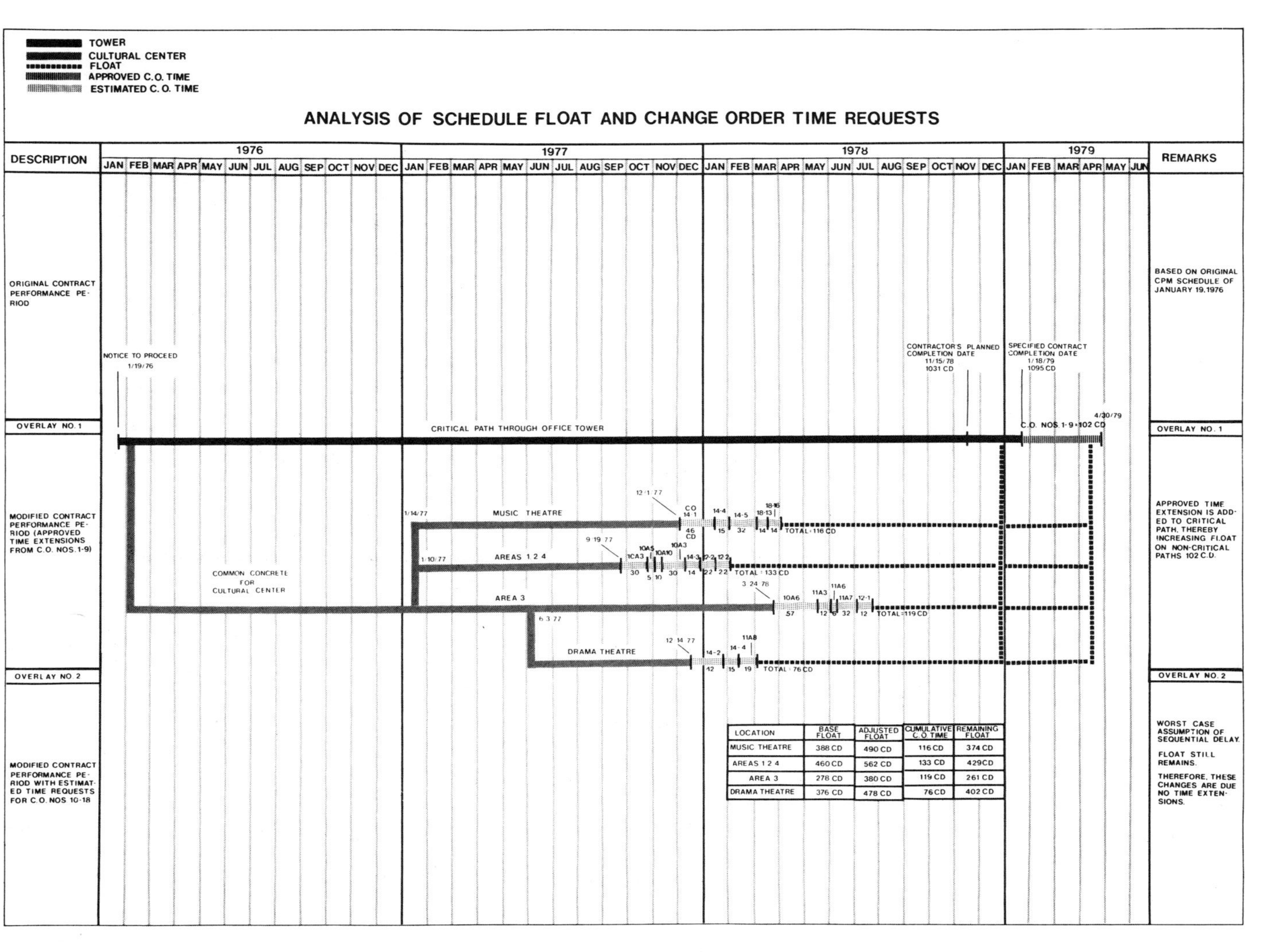

LOCATION	BASE FLOAT	ADJUSTED FLOAT	CUMULATIVE C.O. TIME	REMAINING FLOAT
MUSIC THEATRE	388 CD	490 CD	116 CD	374 CD
AREAS 1 2 4	460 CD	562 CD	133 CD	429CD
AREA 3	278 CD	380 CD	119 CD	261 CD
DRAMA THEATRE	376 CD	478 CD	76CD	402 CD

Figure 8–3: The Effect to the Project Critical Path of Changes

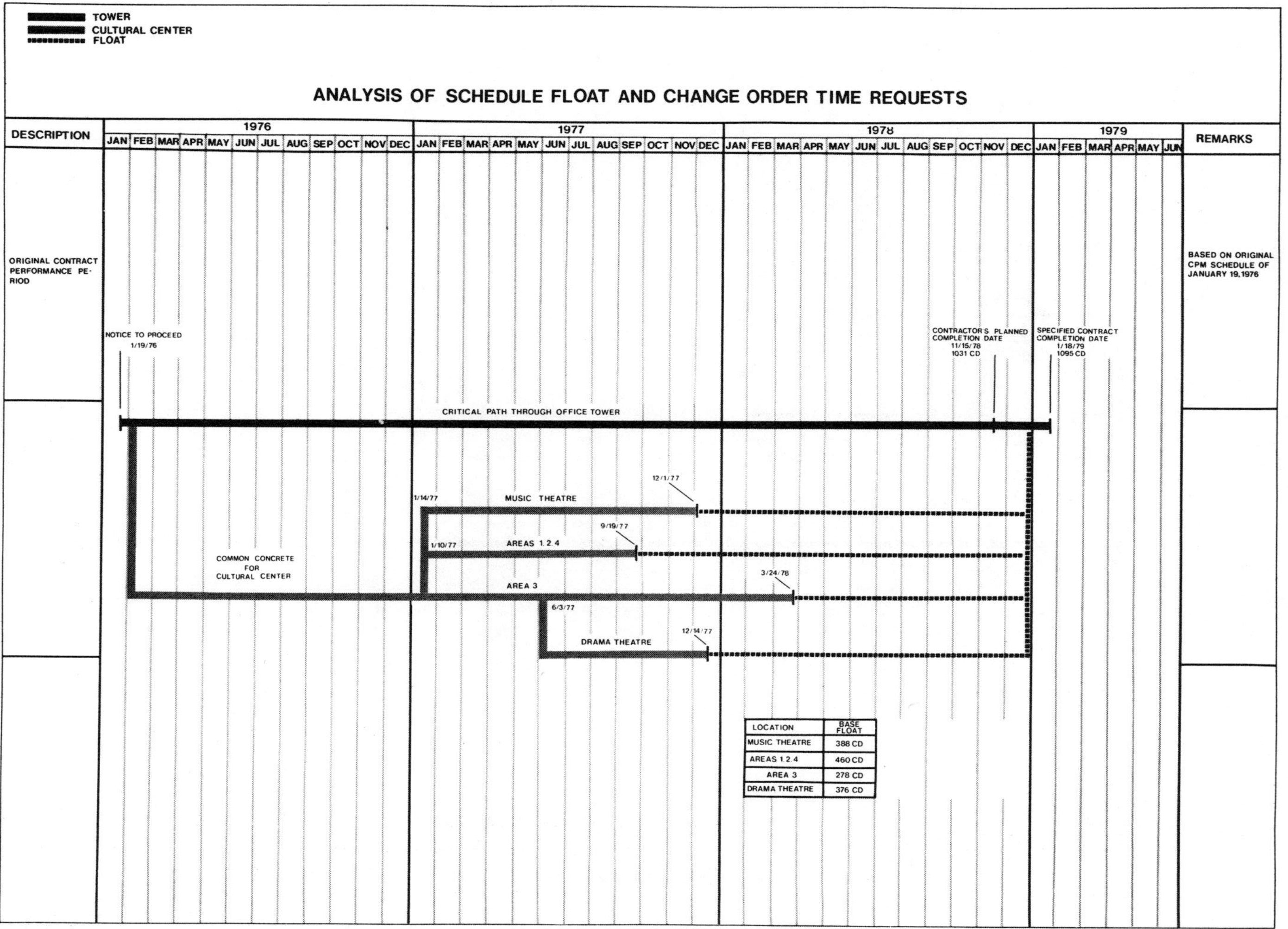

LOCATION	BASE FLOAT
MUSIC THEATRE	388 CD
AREAS 1 2 4	460 CD
AREA 3	278 CD
DRAMA THEATRE	376 CD

Figure 8–4: The Effect to the Project Critical Path of Changes

Conclusion

Hopefully this chapter has increased your awareness of the respective advantages and disadvantages of negotiation, arbitration, and litigation as methods for construction claim resolution. The cost savings, delay avoidance, and control of the proceedings that are possible through negotiation have increasingly prompted otherwise potential litigants to negotiate a mutually satisfactory resolution of disputes.

In those situations where negotiation of claims is unsuccessful or impossible, our discussion of successful methods for presentation of your claims will hopefully aid you not only in dispute resolution but also in everyday corporate operations. Although the techniques and tools of a particular claim analysis may vary, the general considerations underlying the development and use of those methods discussed should form a solid basis on which to build a comprehensive and complete claims presentation or defense.

AMERICAN ARBITRATION ASSOCIATION
CONSTRUCTION INDUSTRY ARBITRATION RULES
(Effective January 1, 1981)

Section 1. AGREEMENT OF PARTIES—The parties shall be deemed to have made these Rules a part of their arbitration agreement whenever they have provided for arbitration under the Construction Industry Arbitration Rules. These Rules and any amendment thereof shall apply in the form obtaining at the time the arbitration is initiated.

Section 2. NAME OF TRIBUNAL—Any Tribunal constituted by the parties for the settlement of their dispute under these Rules shall be called the Construction Industry Arbitration Tribunal, hereinafter called the Tribunal.

Section 3. ADMINISTRATOR—When parties agree to arbitrate under these Rules, or when they provide for arbitration by the American Arbitration Association, hereinafter called AAA, and an arbitration is initiated hereunder, they thereby constitute AAA the administrator of the arbitration. The authority and duties of the administrator are prescribed in the agreement of the parties and in these Rules.

Section 4. DELEGATION OF DUTIES—The duties of the AAA under these Rules may be carried out through Tribunal Administrators, or such other officers or committees as the AAA may direct.

Section 5. NATIONAL PANEL OF ARBITRATORS—In cooperation with the Construction Industry Arbitration Committee, the AAA shall establish and maintain a National Panel of Construction Arbitrators, hereinafter called the Panel, and shall appoint an arbitrator or arbitrators therefrom as hereinafter provided. A neutral arbitrator selected by mutual choice of both parties or their appointees, or appointed by the AAA, is hereinafter called the arbitrator, whereas an arbitrator selected unilaterally by one party is hereinafter called the party-appointed arbitrator. The term arbitrator may hereinafter be used to refer to one arbitrator or to a Tribunal of multiple arbitrators.

Section 6. OFFICE OF TRIBUNAL—The general office of a Tribunal is the headquarters of the AAA, which may, however, assign the administration of an arbitration to any of its Regional Offices.

Section 7. INITIATION UNDER AN ARBITRATION PROVISION IN A CONTRACT—Arbitration under an arbitration provision in a contract shall be initiated in the following manner:

The initiating party shall, within the time specified by the contract, if any, file with the other party a notice of an intention to arbitrate. (Demand), which notice shall contain a statement setting forth the nature of the dispute, the amount involved, and the remedy sought; and shall file two copies of said notice with any Regional Office of the AAA, together with two copies of the arbitration provisions of the contract and the appropriate filing fee as provided in Section 48 hereunder.

The AAA shall give notice of such filing to the other party. A party upon whom the demand for arbitration is made may file an answering statement in duplicate with the AAA within seven days after notice from the AAA, simultaneously sending a copy to the other party. If a monetary claim is made in the answer the appropriate administrative fee provided in the Fee Schedule shall be forwarded to the AAA with the answer. If no answer is filed within the stated time, it will be treated as a denial of the claim. Failure to file an answer shall not operate to delay the arbitration.

Section 8. CHANGE OF CLAIM OR COUNTERCLAIM—After filing of the claim or counterclaim, if either party desires to make any new or different claim or counterclaim, same shall be made in writing and filed with the AAA, and a copy thereof shall be mailed to the other party who shall have a period of seven days from the date of such mailing within which to file an answer with the AAA. However, after the arbitrator is appointed no new or different claim or counterclaim may be submitted without the arbitrator's consent.

Section 9. INITIATION UNDER A SUBMISSION—Parties to any existing dispute may commence an arbitration under these Rules by filing at any Regional Office two (2) copies of a written agreement to arbitrate under these Rules (Submission), signed by the parties. It shall contain a statement of the matter in dispute, the amount of money involved, and the remedy sought, together with the appropriate filing fee as provided in the Fee Schedule.

Section 10. PRE-HEARING CONFERENCE—At the request of the parties or at the discretion of the AAA a pre-hearing conference with the administrator and the parties or their counsel will be scheduled in appropriate cases to arrange for an exchange of information and the stipulation of uncontested facts so as to expedite the arbitration proceedings.

Section 11. FIXING OF LOCALE—The parties may mutually agree on the locale where the arbitration is to be held. If any party requests that the hearing be held in a specific locale and the other party files no objection thereto within seven days after notice of the request is mailed to such party, the locale shall be the one requested. If a party objects to the locale requested by the other party, the AAA shall have power to determine the locale and its decision shall be final and binding.

Section 12. QUALIFICATIONS OF ARBITRATOR—Any arbitrator appointed pursuant to Section 13 or Section 15 shall be neutral, subject to disqualification for the reasons specified in Section 19. If the agreement of the parties names an arbitrator or specifies any other method of appointing an arbitrator, or if the parties specifically agree in writing, such arbitrator shall not be subject to disqualification for said reasons.

Section 13. APPOINTMENT FROM PANEL—If the parties have not appointed an arbitrator and have not provided any other method of appointment, the arbitrator shall be appointed in the following manner: Immediately after the filling of the Demand or Submission, the AAA shall submit simultaneously to each party to the dispute an identical list of names of persons chosen from the Panel. Each party to the dispute shall have seven days from the mailing date in which to cross off any names to which it objects, number the remaining names to indicate the order of preference, and return the list to the AAA. If a party does not return the list within the time specified, all persons named therein shall be deemed acceptable. From among the persons who have been approved on both lists, and in accordance with the designated order of mutual preference, the

AAA shall invite the acceptance of an arbitrator to serve. If the parties fail to agree upon any of the persons named, or if acceptable arbitrators are unable to act, or if for any other reason the appointment cannot be made from the submitted lists, the AAA shall have the power to make the appointment from other members of the Panel without the submission of any additional lists.

Section 14. DIRECT APPOINTMENT BY PARTIES—If the agreement of the parties names an arbitrator or specifies a method of appointing an arbitrator, that designation or method shall be followed. The notice of appointment, with name and address of such arbitrator, shall be filed with the AAA by the appointing party. Upon the request of any such appointing party, the AAA shall submit a list of members of the Panel from which the party may make the appointment.

If the agreement specifies a period of time within which an arbitrator shall be appointed, and any party fails to make such appointment within that period, the AAA shall make the appointment.

If no period of time is specified in the agreement, the AAA shall notify the parties to make the appointment, and if within seven days after mailing of such notice such arbitrator has not been so appointed, the AAA shall make the appointment.

Section 15. APPOINTMENT OF ARBITRATOR BY PARTY-APPOINTED ARBITRATORS—If the parties have appointed their party-appointed arbitrators or if either or both of them have been appointed as provided in Section 14, and have authorized such arbitrator to appoint an arbitrator within a specified time and no appointment is made within such time or any agreed extension thereof, the AAA shall appoint an arbitrator who shall act as Chairperson.

If no period of time is specified for appointment of the third arbitrator and the party-appointed arbitrators do not make the appointment within seven days from the date of the appointment of the last party-appointed arbitrator, the AAA shall appoint the arbitrator who shall act as Chairperson.

If the parties have agreed that their party-appointed arbitrators shall appoint the arbitrator from the Panel, the AAA shall furnish to the party-appointed arbitrators, in the manner prescribed in Section 13, a list selected from the Panel, and the appointment of the arbitrator shall be made as prescribed in such Section.

Section 16. NATIONALITY OF ARBITRATOR IN INTERNATIONAL ARBITRATION—If one of the parties is a national or resident of a country other than the United States, the arbitrator shall, upon the request of either party, be appointed from among the nationals of a country other than that of any of the parties.

Section 17. NUMBER OF ARBITRATORS—If the arbitration agreement does not specify the number of arbitrators, the dispute shall be heard and determined by one arbitrator, unless the AAA, in its discretion, directs that a greater number of arbitrators be appointed.

Section 18. NOTICE TO ARBITRATOR OF APPOINTMENT—Notice of the appointment of the arbitrator, whether mutually appointed by the parties or appointed by the AAA, shall be mailed to the arbitrator by the AAA, together with a copy of these Rules, and the signed acceptance of the arbitrator shall be filed prior to the opening of the first hearing.

Section 19. DISCLOSURE AND CHALLENGE PROCEDURE—A person appointed as neutral arbitrator shall disclose to the AAA any circumstances likely to affect his or her impartiality, including any bias or any financial or personal interest in the result of the arbitration or any past or present relationship with the parties or their counsel. Upon receipt of such information from such arbitrator or other source, the AAA shall communicate such information to the parties, and, if it deems it appropriate to do so, to the arbitrator and others. Thereafter, the AAA shall determine whether the arbitrator should be disqualified and shall inform the parties of its decision, which shall be conclusive.

Section 20. VACANCIES—If any arbitrator should resign, die, withdraw, refuse, be disqualified or be unable to perform the duties of office, the AAA shall, on proof satisfactory to it, declare the office vacant. Vacancies shall be filled in accordance with the applicable provision of these Rules. In the event of a vacancy in a panel of neutral arbitrators, the remaining arbitrator or arbitrators may continue with the hearing and determination of the controversy, unless the parties agree otherwise.

Section 21. TIME AND PLACE—The arbitrator shall fix the time and place for each hearing. The AAA shall mail to each party notice thereof at least five days in advance, unless the parties by mutual agreement waive such notice or modify the terms thereof.

Section 22. REPRESENTATION BY COUNSEL—Any party may be represented by counsel. A party intending to be so represented shall notify the other party and the AAA of the name and address of counsel at least three days prior to the date set for the hearing at which counsel is first to appear. When an arbitration is initiated by counsel, or where an attorney replies for the other party, such notice is deemed to have been given.

Section 23. STENOGRAPHIC RECORD—The AAA shall make the necessary arrangements for the taking of a stenographic record whenever such record is requested by a party. The requesting party or parties shall pay the cost of such record as provided in Section 50.

Section 24. INTERPRETER—The AAA shall make the necessary arrangements for the services of an interpreter upon the request of one or both parties, who shall assume the cost of such services.

Section 25. ATTENDANCE AT HEARINGS—Persons having a direct interest in the arbitration are entitled to attend hearings. The arbitrator shall otherwise have the power to require the retirement of any witness or witnesses during the testimony of other witnesses. It shall be discretionary with the arbitrator to determine the propriety of the attendance of any other persons.

Section 26. ADJOURNMENTS—The arbitrator may adjourn the hearing, and must take such adjournment when all of the parties agree thereto.

Section 27. OATHS—Before proceeding with the first hearing or with the examination of the file, each arbitrator may take an oath of office, and if required by law, shall do so. The arbitrator may require witnesses to testify under oath administered by any duly qualified person or, if required by law or demanded by either party, shall do so.

Section 28. MAJORITY DECISION—Whenever there is more than one arbitrator, all decisions of the arbitrators must be by at least a majority. The award must also be made by at least a majority unless the concurrence of all is expressly required by the arbitration agreement or by law.

Section 29. ORDER OF PROCEEDINGS—A hearing shall be opened by the filing of the oath of the arbitrator, where required, and by the recording of the place, time and date of the hearing, the presence of the arbitrator and parties, and counsel, if any, and by the receipt by the arbitrator of the statement of the claim and answer, if any.

Section 30. ARBITRATION IN THE ABSENCE OF A PARTY—Unless the law provides to the contrary, the arbitration may proceed in the absence of any party, who, after due notice, fails to be present or fails to obtain an adjournment. An award shall not be made solely on the default of a party. The arbitrator shall require the party who is present to submit such evidence as deemed necessary for the making of an award.

Section 31. EVIDENCE—The parties may offer such evidence as they desire and shall produce such additional evidence as the arbitrator may deem necessary to an understanding and determination of the dispute. An arbitrator authorized by law to subpoena witnesses or documents may do so upon the request of any party, or independently. The arbitrator shall be the judge of the admissibility of the evidence offered and conformity to legal rules of evidence shall not be necessary. All evidence shall be taken in the presence of all of the arbitrators and all of the parties, except where any of the parties is absent in default or has waived his or her right to be present.

Section 32. EVIDENCE BY AFFIDAVIT AND FILING OF DOCUMENTS—The arbitrator may receive and consider the evidence of witnesses by affidavit, giving it such weight as seems appropriate after consideration of any objections made to its admission.

All documents not filed with the arbitrator at the hearing, but arranged for at the hearing or subsequently by agreement of the parties, shall be filed with the AAA for transmission to the arbitrator. All parties shall be afforded opportunity to examine such documents.

Section 33. INSPECTION OR INVESTIGATION—An arbitrator finding it necessary to make an inspection or investigation in connection with the arbitration shall direct the AAA to so advise the parties. The arbitrator shall set the time and the AAA shall notify the parties thereof. Any party who so desires may be present at such inspection or investigation. In the event that one or both parties are not present at the inspection or investigation, the arbitrator shall make a verbal or written report to the parties and afford them an opportunity to comment.

Section 34. CONSERVATION OF PROPERTY—The arbitrator may issue such orders as may be deemed necessary to safeguard the property which is the subject matter of the arbitration without prejudice to the rights of the parties or to the final determination of the dispute.

Section 35. CLOSING OF HEARINGS—The arbitrator shall specifically inquire of the parties whether they have any further proofs to offer or witnesses to be heard. Upon receiving negative replies, the arbitrator shall declare the hearings closed and a minute thereof shall be recorded. If briefs are to be filed, the hearings shall be declared closed as of the final date set by the arbitrator for the receipt of briefs. If documents are to be filed as provided in Section 32 and the date set for their receipt is later than that set for the receipt of briefs, the later date shall be the date of closing the hearing. The time limit within which the arbitrator is required to make an award shall commence to run, in the absence of other agreements by the parties, upon the closing of the hearings.

Section 36. REOPENING OF HEARINGS—The hearings may be reopened by the arbitrator at will, or upon application of a party at any time before the award is made. If the reopening of the hearing would prevent the making of the award within the specific time agreed upon by the parties in the contract out of which the controversy has arisen, the matter may not be reopened, unless the parties agree upon the extension of such time limit. When no specific date is fixed in the contract, the arbitrator may reopen the hearings, and the arbitrator shall have thirty days from the closing of the reopened hearings within which to make an award.

Section 37. WAIVER OF ORAL HEARINGS—The parties may provide, by written agreement, for the waiver of oral hearings. If the parties are unable to agree as to the procedure, the AAA shall specify a fair and equitable procedure.

Section 38. WAIVER OF RULES—Any party who proceeds with the arbitration after knowledge that any provision or requirement of these Rules has not been complied with and who fails to state an objection thereto in writing shall be deemed to have waived the right to object.

Section 39. EXTENSIONS OF TIME—The parties may modify any period of time by mutual agreement. The AAA for good cause may extend any period of time established by these Rules, except the time for making the award. The AAA shall notify the parties of any such extension of time and its reason therefor.

Section 40. COMMUNICATION WITH ARBITRATOR AND SERVING OF NOTICES—There shall be no communication between the parties and an arbitrator other than at oral hearings. Any other oral or written communications from the parties to the arbitrator shall be directed to the AAA for transmittal to the arbitrator.

Each party to an agreement which provides for arbitration under these Rules shall be deemed to have consented that any papers, notices or process necessary or proper for the initiation or continuation of an arbitration under these Rules and for any court action in connection therewith or for the entry of judgment on any award made thereunder may be served upon such party by mail addressed to such party or its attorney at the last known address or by personal service, within or without the state wherein the arbitration is to be held (whether such party be within or without the United States of America), provided that reasonable opportunity to be heard with regard thereto has been granted such party.

Section 41. TIME OF AWARD—The award shall be made promptly by the arbitrator and, unless otherwise agreed by the parties, or specified by law, not later than thirty days from the date of closing the hearings, or if oral hearings have been waived, from the date of transmitting the final statements and proofs to the arbitrator.

Section 42. FORM OF AWARD—The award shall be in writing and shall be signed either by the sole arbitrator or by at least a majority if there be more than one. It shall be executed in the manner required by law.

Section 43. SCOPE OF AWARD—The arbitrator may grant any remedy or relief which is just and equitable and within the terms of the agreement of the parties. The arbitrator, in the award, shall assess arbitration fees and expenses as provided in Sections 48 and 50 equally or in favor of any party and, in the event any administrative fees or expenses are due the AAA, in favor of the AAA.

Section 44. AWARD UPON SETTLEMENT—If the parties settle their dispute during the course of the arbitration, the arbitrator, upon their request, may set forth the terms of the agreed settlement in an award.

Section 45. DELIVERY OF AWARD TO PARTIES—Parties shall accept as legal delivery of the award the placing of the award or a true copy thereof in the mail by the AAA, addressed to such party or its attorney at the last known address or by personal service, within or without the state wherein the arbitration is to be held (whether such party be within or without the United States of America), provided that reasonable opportunity to be heard with regard thereto has been granted such party.

Section 46. RELEASE OF DOCUMENTS FOR JUDICIAL PROCEEDINGS—The AAA shall, upon the written request of a party, furnish to such party, at its expense, certified facsimiles of any papers in the AAA's possession that may be required in judicial proceedings relating to the arbitration.

Section 47. APPLICATIONS TO COURT—No judicial proceedings by a party relating to the subject matter of the arbitration shall be deemed a waiver of the party's right to arbitrate.

The AAA is not a necessary party in judicial proceedings relating to the arbitration.

Parties to these Rules shall be deemed to have consented that judgment upon the award rendered by the arbitrator(s) may be entered in any Federal or State Court having jurisdiction thereof.

Section 48. ADMINISTRATIVE FEES—As a nonprofit organization, the AAA shall prescribe an administrative fee schedule and a refund schedule to compensate it for the cost of providing administrative services. The schedule in effect at the time of filing or the time of refund shall be applicable.

The administrative fees shall be advanced by the initiating party or parties in accordance with the administrative fee schedule, subject of final apportionment by the arbitrator in the award.

When a matter is withdrawn or settled, the refund shall be made in accordance with the refund schedule.

The AAA, in the event of extreme hardship on the part of any party, may defer or reduce the administrative fee.

Section 49. FEE WHEN ORAL HEARINGS ARE WAIVED—Where all oral hearings are waived under Section 37 the Administrative Fee Schedule shall apply.

Section 50. EXPENSES—The expenses of witnesses for either side shall be paid by the party producing such witnesses.

The cost of the stenographic record, if any is made, and all transcripts thereof, shall be prorated equally between the parties ordering copies unless they shall otherwise agree and shall be paid for by the responsible parties directly to the reporting agency.

All other expenses of the arbitration, including required traveling and other expenses of the arbitrator and of AAA representatives, and the expenses of any witness or the cost of any proofs produced at the direct request of the arbitrator, shall be borne equally by the parties, unless they agree otherwise, or unless the arbitrator in the award assesses such expenses or any part thereof against any specified party or parties.

Section 51. ARBITRATOR'S FEE—Unless the parties agree to terms of compensation, members of the National Panel of Construction Arbitrators will serve without compensation for the first two days of service.

Thereafter, compensation shall be based upon the amount of service involved and the number of hearings. An appropriate daily rate and other arrangements will be discussed by the administrator with the parties and the arbitrator(s). If the parties fail to agree to the terms of compensation, an appropriate rate shall be established by the AAA, and communicated in writing to the parties.

Any arrangements for the compensation of an arbitrator shall be made through the AAA and not directly by the arbitrator with the parties. The terms of compensation of neutral arbitrators on a Tribunal shall be identical.

Section 52. DEPOSITS—The AAA may require the parties to deposit in advance such sums of money as it deems necessary to defray the expense of the arbitration, including the arbitrator's fee if any, and shall render an accounting to the parties and return any unexpended balance.

Section 53. INTERPRETATION AND APPLICATION OF RULES—The arbitrator shall interpret and apply these Rules insofar as they relate to the arbitrator's powers and duties. When there is more than one arbitrator and a difference arises among them concerning the meaning or application of any such Rules, it shall be decided by a majority vote. If that is unobtainable, either an arbitrator or a party may refer the question to the AAA for final decision. All other Rules shall be interpreted and applied by the AAA.

Administrative Fee Schedule

A filing fee of $150 will be paid at the time the case is initiated.

The balance of the administrative fee of the AAA is based upon the amount of each claim and counterclaim as disclosed when the claim and counterclaim are filed, and is due and payable prior to the notice of appointment of the neutral arbitrator.

In those claims and counterclaims which are not for a monetary amount, an appropriate administrative fee will be determined by the AAA, payable prior to such notice of appointment.

Amount of Claim or Counterclaim	*Fee for Claim or Counterclaim*
Up to $10,000	3% (minimum $150)
$10,000 to $25,000	$300, plus 2% of excess over $10,000
$25,000 to $100,000	$600, plus 1% of excess over $25,000
$100,000 to $200,000	$1,350, plus ½% of excess over $100,000
$200,000 to $5,000,000	$1,850, plus ¼% of excess over $200,000

Other Service Charges

$50.00 payable by a party causing an adjournment of any scheduled hearing;

$100.00 payable by a party causing a second or additional adjournment of any scheduled hearing;

$25.00 payable by each party for each second and subsequent hearing which is either clerked by the AAA or held in a hearing room provided by the AAA.

Refund Schedule

If the AAA is notified that a case has been settled or withdrawn before it mails a notice of appointment to a neutral arbitrator, all of the fee in excess of $150 will be refunded.

If the AAA is notified that a case is settled or withdrawn thereafter but at least 48 hours before the date and time set for the first hearing, one-half of the fee in excess of $150 will be refunded.

Answers to Questions

Chapter Three

1. Set up your project budget.

2. Cost type, area of work, specific activity

3. The project manager

4. Planned versus actual performance

5. Any accident should be reported, no matter how severe.

Chapter Four

1. A major change in scope occurs; actual progress differs significantly from the plan; upper management requests revisions; the planned budget cannot be met.

2. "Cardinal" changes

3. The noncompetitive nature of the pricing

4. Lump sum, unit price, time and material

Chapter Five

1. (a) The bar chart does not show detailed construction activities.
   (b) The bar chart does not show time-consuming off-site activities (e.g., drawing approval, delivery of owner-furnished equipment) which can influence project completion.
   (c) The bar chart does not show the interrelationship between work activities.
   (d) The bar chart is difficult to update or revise.
   (e) The bar chart does not identify the activities critical to project completion.

2. Since the subcontractors generally perform many of the construction activities, they can provide the general contractor with valuable information concerning sequencing and durations of activities for utilization in the overall project schedule. During construction, subcontractors can also provide the general contractor with information concerning potential performance problems (e.g., late approval of electrical fixture submittals) which can have an impact on project completion.

3. Updating the construction schedule is often required by the contract between the parties, and failure to do so may be breach of contract. For the electrical contractor, updating the schedule is the most practical way to identify potential performance problems and to cure them (e.g., request a time extension or a contract modification) and minimize their impact on the time and cost of your performance.

4. The critical path is the longest chain of activities (in terms of time) through the network of activities. Since it is the longest chain of activities in terms of time, any delay in an activity or activities on this chain, if not remedied, will result in a delay of the overall project completion date.

Chapter Six

1. Develop and use your own contract or use a standard contract prepared by a trade association.

2. Size and complexity of the job, duration, location, customer requirements.

3. Normally, the subcontractor need only obtain the permit for his portion of the work.

Index

C

D

E

F

R

S

T

U

V

W